DATE DUE			
Jul 8 '81			

Cellulose Technology Research

Albin F. Turbak, *Editor*

A symposium sponsored by
the Cellulose, Paper, and
Textile Division at the
168th Meeting of the
American Chemical Society,
Atlantic City, N.J.,
September 11–12, 1974

ACS SYMPOSIUM SERIES **10**

AMERICAN CHEMICAL SOCIETY

WASHINGTON, D. C. 1975

Library of Congress CIP Data

Cellulose technology research.
(ACS symposium series; 10)

Includes bibliographical references and index.

1. Cellulose—Congresses.

I. Turbak, Albin F., 1929- ed. II. American Chemical Society. Cellulose, Paper, and Textile Division. III. Series: American Chemical Society. ACS symposium series; 10.

TS933.C4C43 674'.134 75-2021
ISBN 0-8412-0248-6 ACSMC8 10 1-353

PRINTED IN THE UNITED STATES OF AMERICA

ACS Symposium Series

Robert F. Gould, *Series Editor*

FOREWORD

The ACS Symposium Series was founded in 1974 to provide a medium for publishing symposia quickly in book form. The format of the Series parallels that of its predecessor, Advances in Chemistry Series, except that in order to save time the papers are not typeset but are reproduced as they are submitted by the authors in camera-ready form. As a further means of saving time, the papers are not edited or reviewed except by the symposium chairman, who becomes editor of the book. Papers published in the ACS Symposium Series are original contributions not published elsewhere in whole or major part and include reports of research as well as reviews since symposia may embrace both types of presentation.

CONTENTS

PREFACE

Recent oil shortages have pronounced an end to the era of cheap petrochemicals. The current furor over vinyl chloride toxicity coupled with the biodegradability problems of many petrochemical-based materials finally have created a long overdue awareness that synthetics may have very serious shortcomings and may not represent the total panacea which large oil interests and their Madison Avenue marketing agencies have incessantly claimed over the years.

These cost and health factors have aroused a renewed appreciation for natural-product materials. As a result, cellulose and cellulose-based products are presently enjoying an increased research impetus. Cellulose is without a doubt not only the most abundant polymer in the world but also the most versatile. Its present commercial use in manufacturing artificial kidneys, aspirins, hot dogs, ice cream, paints, diapers, contact lens lubricants, fibers, plastics, and a host of other products clearly demonstrates that it can be adapted by creative scientists to serve practically any desired function.

This symposium was held in an effort to present a cross section of recent research in the cellulose area. Judging by the scope of these efforts, much more work is undoubtedly underway which is not presently available for public disclosure. The contributions in this volume, however, adequately demonstrate that cellulose researchers are diligently pursuing efforts to contribute basic knowledge and to capitalize on the opportunities being created by the present dynamic market needs and conditions. Each author is individually responsible for his overall claims since it is ACS policy for this series that the symposium chairman not review the final submissions in order to expedite publication.

I would like to thank each of the participants for their kind cooperation, and I hope that this joint effort proves useful and stimulating to fellow researchers.

ITT Rayonier, Inc.
Whippany, N.J.
Dec. 23, 1974

ALBIN F. TURBAK

1

Colloidal Microcrystalline Celluloses

O. A. BATTISTA

Research Services Corp., 5280 Trail Lake Drive, Fort Worth, Tex. 76133

The science and technology of polymers during the past few decades have been centered largely around the phenomena and the products that result from precipitating or "freezing" long chain molecules into fixed matrices leading to important commercial structural shapes such as fibers, films, plastics, and coatings.

The basic hypothesis around which the science of microcrystalline colloidal polymer chemistry (1) is unfolding necessitates that specific requisites must be met; it is in the deliberate combination of these requisites that it derives its value and originality.

Firstly, the molecular weight of the individual long-chain molecules must be high enough; the chain molecules must be long enough to crystallize out of solution or from a melt into a two-phase network structure comprising regions of high lateral order (or crystallinity) and regions of low lateral order (or low crystallinity). Secondly, a pretreatment must be involved which is capable of unhinging or loosening the individual microcrystals within their precursor matrix without excessively swelling them or destroying their "crystallinity." Thirdly, once the individual microcrystals have been properly unhinged or loosened within the polymer matrix, they must next be freed by the proper kind of mechanical energy. The individual microcrystals, comprising as they do hundreds of long-chain molecules aggregated together, will now act as discrete, independent, submicron colloidal particles.

Colloidal microcrystalline celluloses were the first of a family of new products that have emerged in recent years (Table 1).

The size and shape of natural cellulose microcrystals are more or less fixed by nature. But wide variations in the dimension of such particles are possible by the appropriate choice of the natural or synthetic precursor raw material (2,3,4).

Figures 1 and 2 are electron micrographs illustrating the

Table 1

Nine Members of the Microcrystal Polymer Products Family.

1. AVICELS....................FROM CELLULOSES
2. AVIAMYLOSES...............FROM AMYLOSES
3. AVIBESTS..................FROM CHRYSOTILES
4. AVITENES..................FROM COLLAGENS
5. AVIAMIDES.................FROM NYLONS
6. AVIESTERS.................FROM POLYESTERS
7. AVIOLEFINS................FROM POLYPROPYLENES
8. AVISILKS..................FROM NATURAL SILKS
9. AVIWOOLS..................FROM NATURAL WOOLS

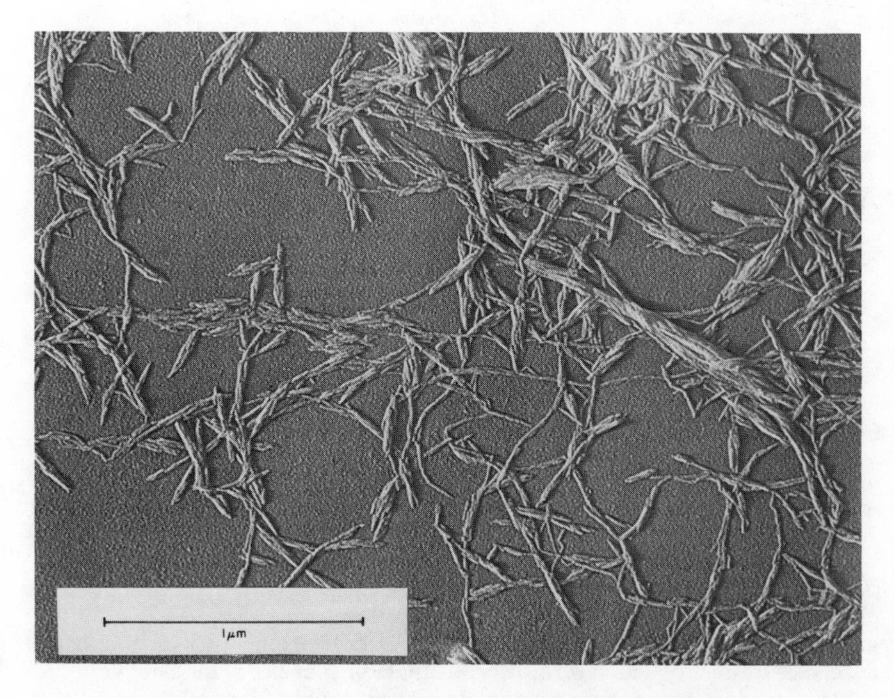

Figure 1. Microcrystals from wood pulp alpha cellulose

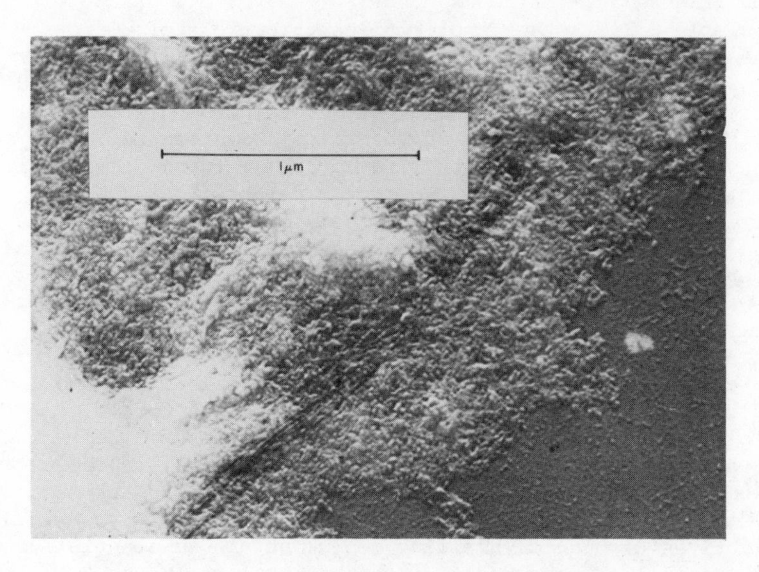

Figure 2. Microcrystals from rayon tire cord

wide spread in the lengths of cellulose microcrystals that is possible; microcrystals of wood cellulose are compared with micro-crystals from viscose rayon at the same magnification.

Cellulose is highly crystalline, uniquely isotactic linear polymer. It contains microcrystals hinged together by true (covalent) molecular bonds. Mechanical beating of cellulose fibers leads only to a separation of the fibrils, but mechanical energy alone cannot break the 1,4 β-glycosidic bonds in the molecular chains going from one crystallite to the other within the fibril with any degree of effectiveness. However, once these hinges in the interconnecting areas are broken, in the case of cellulose by the use of the hydronium ion (HCl), then mechanical energy can be used to cause the individual unhinged microcrystals to disperse into a liquid medium as individual colloidal particles. Not until a sufficiently large number of microcrystals is so freed does a mechanical gel network develop to give a smooth, lard-like gel. Figure 3 illustrates the nature of such aqueous suspensoids for microcrystalline cellulose and for five other members of the microcrystalline polymer family.

The microcrystals of pure cellulose in aqueous gels do not melt, of course. This gives such suspensoids a unique functional property which makes possible the development of a new family of convenience foods; microcrystalline cellulose gels are used in precooked cans of tuna fish, ham, chicken, turkey, and even potato salads as heat-insensitive salad dressings. No previously known edible salad dressing formulation could stand up under the severe sterilization requirements for such canned foods (Figure 4).

Table 2 lists the functional contributions of microcrystalline celluloses in various product uses.

Figure 5 illustrates the mechanism whereby dry colloidal particles of crystalline cellulose, containing numerous "holes" varying from 10 A-100 A in diameter, may be produced. This new porous form of highly crystalline cellulose in powder form is capable of absorbing oils, greases, catalysts, etc. When a dilute slurry of a suspension of individual microcrystals and large aggregates of unhinged microcrystals is spray-dried under proper conditions, the free microcrystals reagglomerate into man-made clusters not unlike the manner in which wooden matchsticks aggregate when piled on a table top.

Still other fascinating opportunities present themselves when chemistry is wedded to these novel colloidal macromolecular particles. For example, reaction of the microcrystalline cellulose crystals proceeds with particular ease and speed. Derivatives can be formed which are also colloidal. These are entirely new materials with very different properties and potential applications. At high degrees of substitution (D.S.), derivatives of microcrystalline cellulose are substantially the same material as produced from conventional cellulose. At low

Figure 3. Aqueous suspensoids of six classes of microcrystal polymer products

Figure 4. Typical commercial food products containing microcrystalline cellulose

Table 2

Funtional Properties and End Uses of Microcrystalline Celluloses

Functionality	Product Uses
Emulsion Stability at High Temperatures	Heat Stable Dressings
Ice-Crystal Control Stability To Effects of Heat Shock Improved Body and Texture	Frozen Desserts
Foam Stability Freeze/Thaw Stability Improved Body and Texture	Whipped Toppings
Gelling Agent	Low Calorie Dressings
Thickener	Sauces and Gravies
Suspending Agent	Suspension of Food Solids
Non-Nutritive Filler	Confections Baked Goods

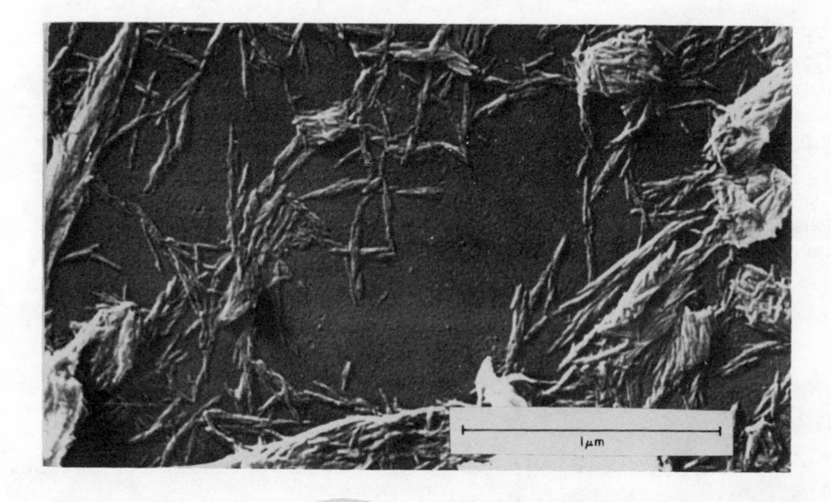

Figure 5. Reaggregated cellulose microcrystals as porous colloidal particles

degrees of substitution, where the colloidal nature of the microcrystals is maintained, the derivatives form unique colloidal dispersions. Dispersions of at least 20% solids in water can be produced. These may have the appearance of greases, ointments, or lotions, depending on the extent of topochemical derivatization and the nature of the groups added.

We have described microcrystalline celluloses, this first in a new species of colloidal microcrystal polymer products to reach world-wide commercial success. The knowledge gleaned from converting fibrous celluloses into new, useful colloidal forms has guided us into converting other linear and crystallin polymer precursors into useful colloidal forms. To date these have included: colloidal microcrystalline amyloses, microcrystalline mineral silicates, microcrystalline collagens, microcrystalline polyamides, microcrystalline polysters, microcrystalline polypropylenes, microcrystalline silks, and microcrystalline wools. A treatise encompassing the results of our research in the newest fields of Microcrystal Polymer Science in collaboration with many associates over a period of 20 years now is available (1). This author predicts that numerous new avenues of opportunity remain unexplored for the more recent members of the microcrystal polymer science family.

Literature Cited

1. Battista, O.A., Microcrystal Polymer Science, A Treatise. McGraw-Hill Book Co., Inc. 1975.

2. Battista, O. A., and Smith, P.A., Ind. Eng. Chem. (1962) 54, (9), 20-29.

3. Battista, O.A., et al, Ind. Eng. Chem., (1956) 48 333-335.

4. Battista, O.A., and Smith, P.A., U.S. Patent No. 2,978,446 (Level-off D.P. Cellulose Products), 1961, April 4.

2

Continuous Thiocarbonate–Redox Grafting on Cellulosic Substrates

W. JAMES BRICKMAN

Scott Paper Co., Scott Plaza III, Philadelphia, Pa. 19113

Abstract

Permanent fire retardancy, as well as other properties, have been grafted onto moving cellulosic substrates as continuous operations. Although earlier use of batch grafting required reaction times of 15 to 120 minutes, continuous grafting has been reduced to less than one to three minutes, under conditions which have been designed for use in textile mills. Example reagent concentrations, formulations, reaction conditions and product properties illustrate the practicability of the continuous thiocarbonate-redox grafting reactions to produce desired modifications on cellulosic materials. Fire retardant grafts which imparted improved abrasion resistance, enhanced chemical resistance (to chlorine bleaches, mildew, bacterial attack, etc.), have appeared to be superior to other known methods of making cotton fire retardant. The use of microwave energy and the new monomer, diethylphosphatoethyl methacrylate, have had their roles revealed in production of fire retardancy by the continuous grafting process.

Introduction

Methods for batch graft copolymerization of vinyl monomers on to cellulosic materials by the thiocarbonate-redox method were described earlier (1-3). Faessinger and Conte discovered that mildly xanthated cellulose, in the presence of vinyl monomers and oxidizing catalysts, could be graft copolymerized to form a true graft copolymer (4-12). This discovery was first revealed in Belgian patents (4,5) in 1964, followed by a series of other patents (6-12) which may have led to some of the publications of other investigators in the field. Amongst the foremost of these have been our colleagues in Austria, H.A. Kraessig and his associates (3,13-15), whose papers have dealt primarily with reaction mechanisms and product properties of viscose rayon grafts. Authors in other parts of the world (16-23) have reported experimentation conditions which, in many cases, have used batch techniques similar to those given in the original patent examples. When such extreme conditions are employed, i.e. xanthation to gamma values of 6 to 67, combined with peroxide concentrations up to 10% owf (on weight of fiber), degradation of physical properties, homopolymer formation and other problems have resulted, which have made such methods unsuitable for either batch or continuous grafting of textiles. As a result, certain generalizations should be viewed sceptically, since they only apply in the regions of the actual experimentation. They do not apply to either the batch conditions which we have previously reported(1-3), nor do they apply for the continuous grafting conditions given in this paper.

This presentation will follow the general format:

CONTINUOUS THIOCARBONATE-REDOX GRAFTING ON CELLULOSIC SUBSTRATES

1. BACKGROUND INTRODUCTION
2. MICROWAVE AND ITS UTILIZATION
3. CONTINUOUS PROCESS OUTLINE
4. TYPICAL THIOCARBONATION CONDITIONS
5. POST-THIOCARBONATION TREATMENTS
6. TYPICAL MONOMER FORMULATIONS (a) Fire Retardant
   (b) Water Dispersible
7. "PEPM" DESCRIPTION
8. GRAFTED PRODUCT PROPERTIES (a) Fire Retardancy
   (b) Abrasion Resistance
   (c) Rot Resistance
   (d) Dispersibility & Ion Exchange
9. SUMMARY

Several years ago, we made the interesting discovery that the graft copolymerization step could be shortened from the 15 to 120 minutes range, down to the range of from 3 to 30 seconds. This was achieved by moving monomer-catalyst impregnated, thiocarbonated, cellulosic substrates through a microwave (MW) energy applicator (24-25). Since this discovery was our key to the development of a continuous grafting process at accelerated speeds, we will give a very brief description of microwaves and their application.

Microwave and Its Utilization

In conventional heating systems which involve conduction, induction, or infrared radiation, the highly activated molecules adjacent to the energy sources must transfer their energy to their neighboring molecules through high speed collisions. Initial temperature differentials cause thermal gradients and a finite time is required to uniformly heat the whole mass to reaction temperatures. Microwaves, on the other hand, simultaneously excite all the molecules in a mass which are capable of forming dipoles, thereby avoiding thermal gradients and the attendant migration problems.

The F.C.C. has set aside seven I.S.M. frequencies outside those used for communications. These are:

For INDUSTRIAL, SCIENTIFIC and MEDICAL (I.S.M) Use[●]

	FREQUENCIES MHz.	WAVELENGTHS cm.
MACROWAVE	13.56	2,200
	27.12	1,100
	40.68	740
MICROWAVE	915	33
	2,450	12.5
	5,800	5.1
	22,125	1.35

● - ASSIGNED BY F.C.C.

The following discussion will be concerned with 2,450 MHz frequency microwave and the effects on our aqueous reaction system.

Microwaves fall between macrowaves and infrared regions on the electromagnetic spectrum, with wavelengths of approximately three millimeters to three meters and frequencies ranging from

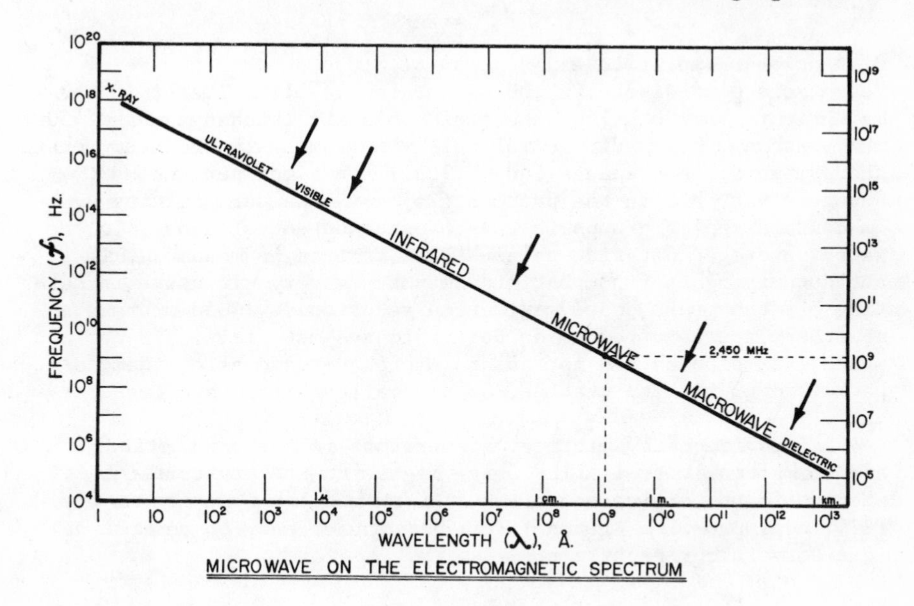

MICROWAVE ON THE ELECTROMAGNETIC SPECTRUM

ca. one hundred million to one hundred billion Hertz (cycles per second). Induction heating usually occurs in conductors, where eddy currents of frequencies not exceeding one Megahertz are used. However, the term "dielectric heating" is applied to use of electromagnetic waves of from one to 150 MHz. Infrared heating is most common for conduction heating and encompasses wavelengths from about 3 microns to 3 millimeters in length, with corresponding one hundred billion to 100 trillion cycles per second frequencies.

ENERGY TRANSFER REGIONS

	FREQUENCY f		WAVELENGTH λ	
	Hz.	MHz.	cm.	m.
INFRARED	10^{14} to 10^{11}	10^{8} to 10^{5}	0·000,3 to 0·3	0·000,000,3 to 0·003
MICROWAVE	10^{11} to 10^{8}	10^{5} to 10^{2}	0·3 to 300	0·003 to 3·0
MACROWAVE	10^{8} to 10^{6}	10^{2} to 1	300 to 30,000	3·0 to 300
DIELECTRIC (INDUCTION)	$<10^{6}$	<1		>300

$$f\lambda = 2{\cdot}99793 \times 10^{10} \text{ cm./sec.}$$

An oversimplified explanation of microwave heating is: "each molecular dipole in the mass tries to align itself in the magnetic field, but since the field polarity is changing at 2450 MHz, each and every dipole molecule simultaneously tries to rotate about its axis and change ends 4.9 billion times per second", - and gets very hot in the process (26)! Since our grafting system contains polar groups in the thiocarbonated cellulose, in the water, hydrogen peroxide catalyst, as well as in monomers, it is not surprising to find that microwaves are very effective in setting off the graft copolymerization reactions. Good microwave absorbers are referred to as being "lossy" materials. Materials which lack dipoles, do not absorb microwaves and allow them to pass through without heating, so are called "microwave transparent".

Fixed frequency microwave generators for 915 and 2450 MHz are commercially available, using magnetron and klystron tubes. We have found Gerling Moore's 2.5 Kw MW modular type generators (27), equipped with Litton Industries magnetrons, to be most satisfactory in our experiments.

One of the interesting properties of microwaves is that they can be conducted, or flow, through rectangular metal pipes or waveguides, when these are made to certain dimensions with fixed relationships to microwave wavelengths. Microwaves are therefore conducted from generators, through waveguides, to microwave applicators. Applicators fall into three general classes:

1. Slotted Waveguides; 2. Fringing Fields; 3. Resonant Cavities. Each type of applicator has special characteristics which make it uniquely suitable for specific substrate configurations. Although we have worked with all three types, most of the following discussions are in terms of the resonant cavities, since these have better characteristics for handling wide and/or thick substrates. The widely advertised "Radar Ovens" and "Microwave Ovens" for home use, are of the resonant cavity type. Commercial, continuous microwave chicken cooking ovens, which handle loads similiar to some of our projected grafting loads, have already been successfully operated for a number of years.

MICROWAVE SUMMARY

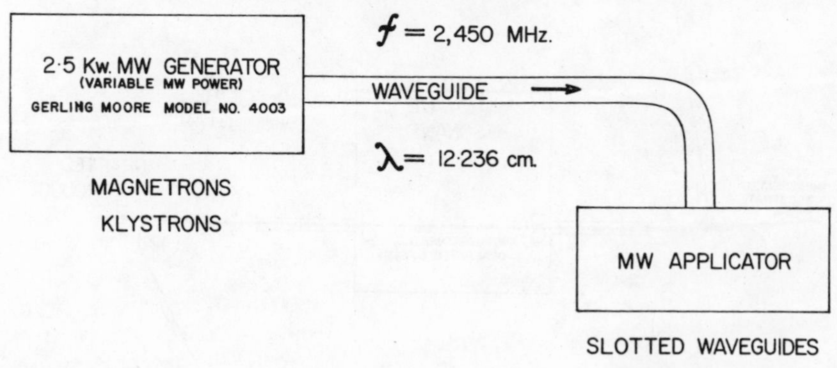

"LOSSY" - materials absorbing MW

"TRANSPARENT" - materials which allow MW to pass without absorbing

Conversion of alternating current at 50-60 Hz, to microwave at 2450 MHz, is approximately 50% efficient. The "coupling efficiency", or conversion of microwave energy to useful work, can be above 80%. In order to estimate the microwave generator requirements, the following equation is useful.

$$\text{MW, Kw.Hr.} = \frac{1}{A} \times \frac{\Delta T, {}^{\circ}C.}{860} \times \text{LOAD}^{*}$$

A - EFFICIENCY OF MW CONVERSION TO USEFUL ENERGY (ca. 80%)

* - $\sum$(LOAD COMPONENTS, Kg./Hr. X SPECIFIC HEATS)

Once the monomer-catalyst system has been applied to the activated substrate, the thiocarbonate-redox grafting system becomes very sensitive to the presence of gases (particularly air) contacting the substrate surfaces. Immediate sealing of the surfaces from contact with the air, through use of a pair of microwave transparent, continuous belts to form an envelope over the substrate, has proved to be a successful solution to this problem. Teflon-coated fiberglas fabrics have been found to be most satisfactory as belt materials. The monomer-catalyst impregnated, thiocarbonated, cellulosic substrate is passed through the resonant cavity microwave field, sealed between the two vapor impervious belts, to kick off the grafting reaction uniformly throughout the substrate. The use of deaerated, saturated steam in the microwave cavity has proved advantageous, largely to offset radiant energy losses through the belt surfaces and to maintain the belts at reaction temperatures.

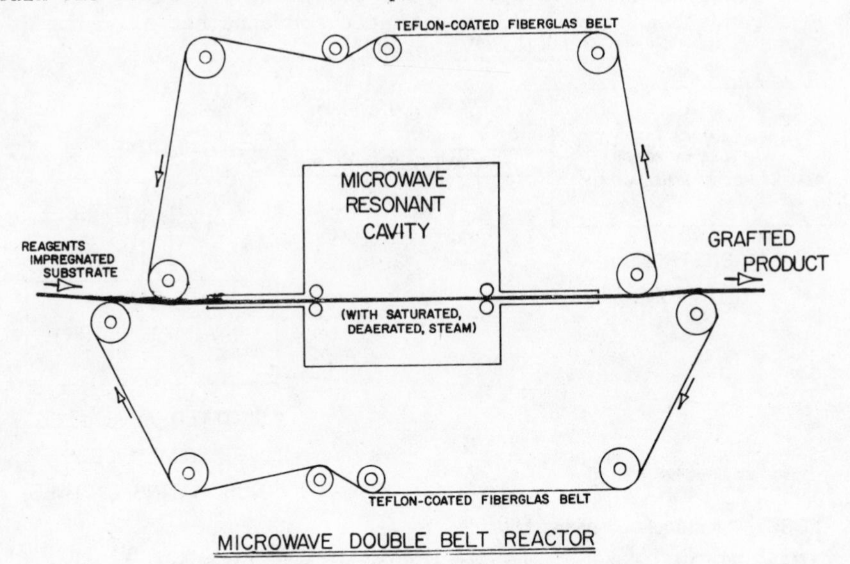

MICROWAVE DOUBLE BELT REACTOR

When the monomer-catalyst impregnated substrates are passed through the microwave field with no envelope of belts, it has been observed that the fabric surfaces (in contact with air, or steam) are not grafted, although excellent grafting occurs throughout the center of the cross-section. Once grafting has been initiated with microwave, sealed between the belts, direct contact with the deaerated, saturated steam on the surfaces of the substrate may be used to complete reactions and maximize grafting efficiencies.

When deaerated, saturated steam is used without benefit of microwaves and double belt reactor, surface grafting appears to be inhibited, even on thin substrates. Also, slightly longer reaction times are usually required, with some reductions in the grafting efficiencies. On thicker substrates (fiber batts, blankets, or fleeces, etc.), temperature gradients and slower heat transfer rates allow thermal migration to occur, with the attendent non-uniformities in grafting. Microwave obviates these problems.

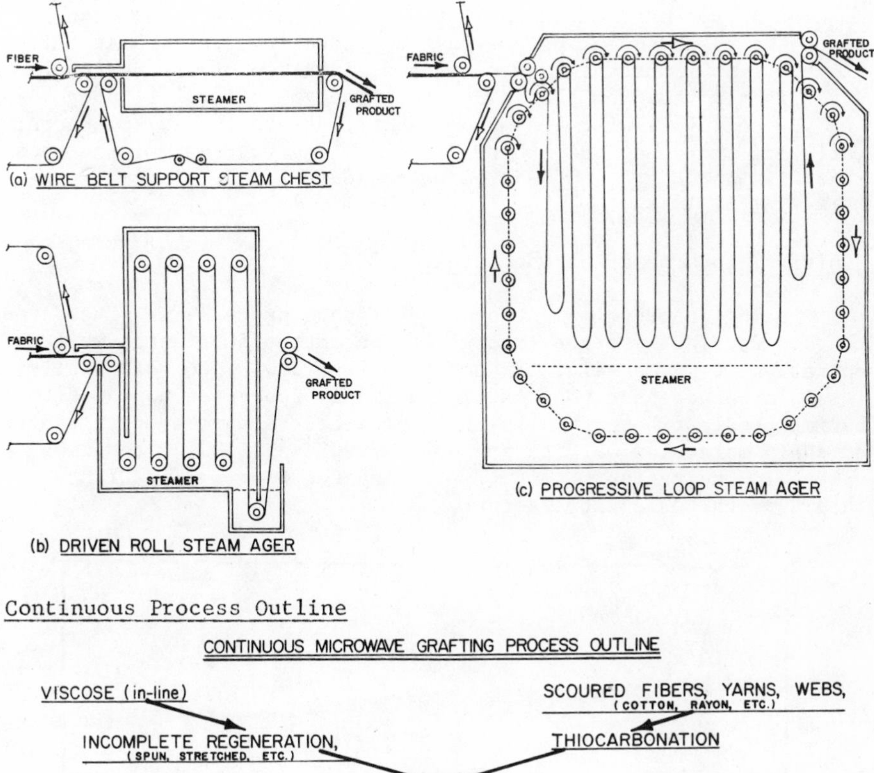

Continuous Process Outline

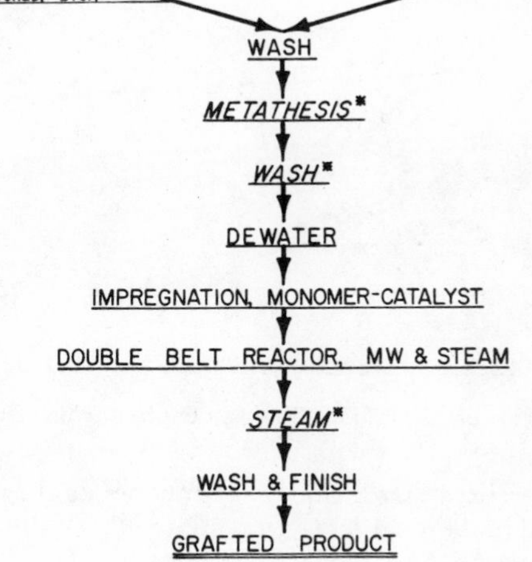

Once the desired levels of thiocarbonation, or "gamma value" have been obtained, whether through the incomplete regeneration of viscose, or via thiocarbonation of fully regenerated rayons, cottons, or other cellulosic materials, all excess reagents and soluble byproducts are washed out of the substrates. The inclusion of an optional metathesis step, followed by a final wash, reduces the subsequent polymerization induction times. At this stage, it is important that as much as possible of the residual water is removed from the substrate prior to uniformly impregnating the cellulose with the monomer-catalyst systems. Immediately following impregnation with the monomer-catalyst system, the substrate is sealed between the two belts and reacted. Normal textile finishing can then be carried out.

Typical Thiocarbonation Conditions

During the development of a continuous process, we discovered that we were able to get good grafts on cotton substrates while operating at gamma values as low as 0.3 to 1.0. Coincidentally, it was observed that the textile physical properties appeared to be improved. Theoretically it is possible to have average grafted chain molecular weights as low as 7,000 to 14,000 with gamma values in the region of 0.6. For many reaction applications, this appears to be satisfactory.

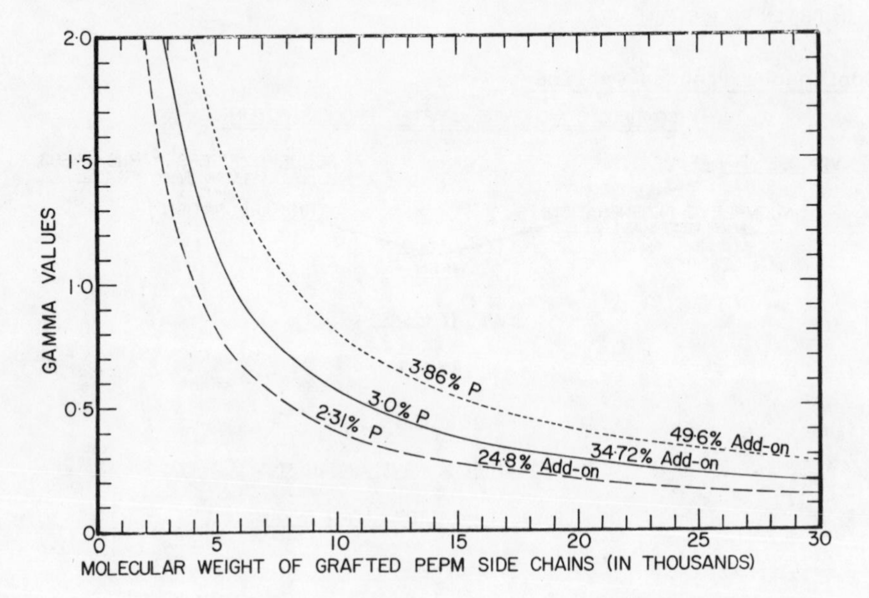

Reaction times, temperatures and reagent concentrations in thiocarbonation are adjusted to fit the substrate conditions. Typical conditions for cotton fibers and cotton fabrics are shown in the following table.

TYPICAL COTTON THIOCARBONATION CONDITIONS

COTTON FIBER BATTS		REAGENTS	COTTON FABRICS	
RATIOS	Moles/AHGU*		Moles/AHGU*	RATIOS
83	0·5-0·7	SODIUM HYDROXIDE	0·3-0·5	50
2		TRITON QS-30		2
120	0·4-0·6	CARBON DISULFIDE	0·2-0·4	60
5		TRITON X-155		5
1		TRITON X-200		1
1		SANDOPAN DTC		1
1,400		WATER		1,667
		CONDITIONS		
300 - 370		% WET PICK-UP	200 - 300	
3 to 5 at 15 to 25		MINUTES at °C.	2 to 4 at 35 to 38	

* — ANHYDROGLUCOSE UNITS

The substrates should be scoured and free from waxes, or finishes which might interfere with the penetration of the reagents. The use of a J-box to give residence time at temperatures above ambient, has been found to be useful. Fully regenerated rayons normally require somewhat milder thiocarbonation conditions, due to their greater accessibility to reagents.

The controlled, incomplete regeneration of viscose rayon during spinning, stretching and washing, is part of the art developed by the individual rayon producers. The techniques vary from plant to plant.

Post-thiocarbonation Treatments

Washing, after thiocarbonation, whether with in-line rayon, or re-xanthated rayon or cotton, has the same objective, i.e. complete removal of all water-soluble by-products, excess reagents or other impurities. Since it is necessary to remove all the free alkali (or acids) on the substrates, without over-washing, measurement of the equilibrium pH of each wash liquor stage is used as a check on the progress of washing after thiocarbonation (or after partial regeneration). Since the efficiency of washing will depend on the design of the washer and on the substrate being processed, we find that washing to a final pH 7.0 ± 0.5 is the safest basis for prediction; particularly when the water supply pH may vary widely.

Once post-thiocarbonation washing (or post-regeneration washing) has been completed, the optional metathesis step may be carried out as a penultimate wash step in which ca. one to three parts per million of ferrous ions are included in a wash water through the addition of Mohr's Salt (ferrous ammonium sulfate hexahydrate). The objective is to have not more than 20 to 30 ppm. Fe^{++} on the thiocarbonated activated cellulose. This is equivalent to less than 30% of the thiocarbonated sites at a gamma value as low as 0.3, and serves only to shorten the induction time for some monomers in the subsequent polymerization. A final wash to remove excess reagents completes the metathesis operation.

POST-XANTHATION WASHING

AFTER THIOCARBONATION AFTER INCOMPLETE REGENERATION

7·0-7·5 EQUILIBRIUM FINAL WASH pH 6·5-7·0

METATHESIS (WASH)

0·002% MOHR'S SALT (ca. 3 ppm. Fe at pH 6·5-7·0)

POST-METATHESIS WASHING

SINGLE RINSE TO REMOVE ALL SOLUBLE REAGENTS

DEWATERING
TO

After Thiocarbonation		After Incomplete Regeneration
28-38	% WATER CONTENT	28-52
39-61	% WET PICK-UP	39-108

The removal of as much of the excess water as possible from the substrate, prior to impregnation with the monomer-catalyst system, is most important. Uniform "wet-on-wet" reagent applications, particularly of emulsions, requires that minimum opportunity be provided for residual water to either block the rapid penetration of monomer-catalyst into the substrate, or to be exchanged by the liquors through preferential absorption on the cellulose by the higher viscosity liquor components. Our experience has shown that best results are usually obtained when the residual water in lightweight cotton fabrics (i.e. less than 3 oz./yd^2.), does not exceed 40 to 50% Wpu (Wet pick-up). Medium weight cotton fabrics (3 to 7 oz./yd^2.) may have 60 to 75% Wpu, while cotton or rayon fiber blankets ("batts" or "fleeces") should be dewatered to ca. 60 to 120% Wpu, depending somewhat on the substrate thicknesses. It should be kept in mind that use of heat to partially dry the substrate is not desirable, since it will also lead to premature decomposition of the active thiocarbonated sites. While grafting with certain systems can be successfully carried out on substrates with higher residual water contents (higher % Wpu), it is usually easier to achieve uniformity and/or high add-ons when residual water is at a minimum prior to impregnation with the monomer-catalyst system.

Typical Monomer Formulations

Monomer-catalyst systems are handled either in the form of solutions, or oil-in-water, or water-in-oil emulsions, depending on the concentrations required and the water-monomer mutual solubility properties. The following typical system, employing the fire retardant monomer, "PEPM", is usually mixed in two parts; the aqueous phase consisting of water, hydrogen peroxide and water solution of CYANAMER P-250 (viscosity builder) is emulsified just before use in the oil phase, which consists of a solution of the other components in the PEPM.

TYPICAL MONOMER-CATALYST SYSTEM

% ows	REAGENT	% owf
43·04	PEPM	41·97
9·71	SOLVENTS IN MONOMER	9·46
1·0	SPAN 85	0·98
1·0	ATLOX G-3409F	0·98
0·025	CYANAMER P-250	0·024
0·61	HYDROGEN PEROXIDE	0·60
44·61	WATER	43·50
100	TOTALS	97·51

% OWS = PERCENT ON WEIGHT OF SYSTEM (SOLUTION OR EMULSION)
% OWf = PERCENT ON WEIGHT OF FABRIC (OR FIBER) = % Wpu

When uniformly applied to scoured and bleached, 3.7 oz./yd^2., cotton flannelette at 97.5% Wpu, the product averaged 38.0% weight increase, with ca. 91% conversion of monomer, even when microwave was omitted and steam was used in the double belt reactor microwave chamber and in a final steam chamber. The product passed the DOC FF 3-71 tests after 50 HLTD (Home Laundering Tumble Drying) cycles.

TYPICAL WATER DISPERSIBLE FIBER MONOMER SYSTEM

% ows	REAGENTS	% owf
0·252	ETHYLENEGLYCOL DIMETHACRYLATE	0·251
66·415	METHACRYLIC ACID	66·082
1·200	HYDROGEN PEROXIDE	1·194
32·133	WATER	31·972
100	TOTAL	99·499

Rayon cellulose fibers of textile lengths do not normally disperse in water, as would the much shorter wood pulp fibers used in paper making. Instead, the longer fibers clump together and form stringy lumps. However, when grafted with monomer systems containing carboxylic acid goups, textile length fibers can be made water dispersible and can be "wet laid" into non-woven webs. Acrylic and/or methacrylic acids, applied to the thiocarbonated, activated substrates at ca. 100% Wpu, have used the above formulation. For example, when microwave was used, combined with steam in the microwave chamber, approximately 54% weight increase was achieved, corresponding to 82% monomer conversion and a neutralization equivalent of about 4 meq./g. (milliequivalents per gram) on cellulosic fibers processed in batt form. The dispersibility of grafted rayon fibers in water has been reported previously (1,3,15), so need not be repeated.

Once the cellulose substrate has been grafted, then the new polymer chains are permanently attached to the fibers. This permits all normal washing, dyeing, or other textile finishing procedures to be carried out on the grafted products.

Reaction times ranged from about one to three minutes in our pilot plant line. In mill trials, we have had good results with as low as 55 seconds in a steam ager, but we would prefer to formulate for over 90 seconds minimum reaction time.

"PEPM" Description

Government regulations, passed and pending, have led to an avalanche of publications on the various aspects of fire retardancy. They were also the driving force behind our search for the best monomer to impart grafted fire retardancy to cellulosic

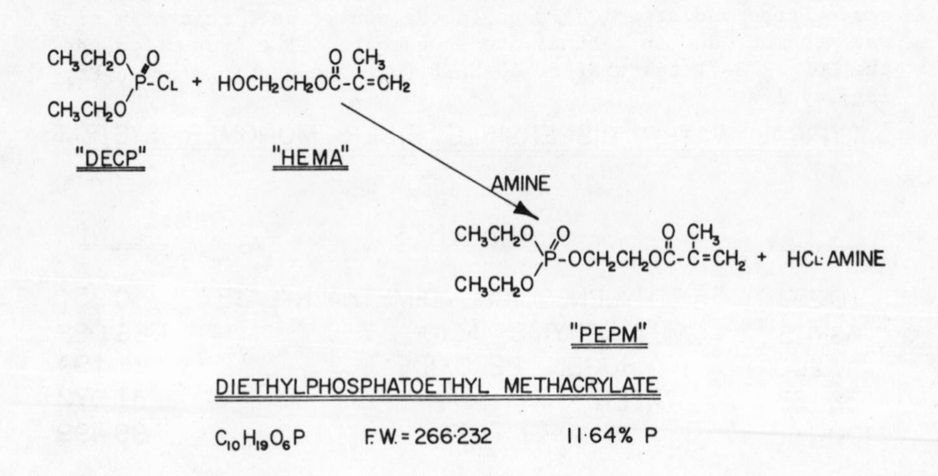

substrate materials, particularly cotton. After screening many compounds, it has been discovered that the one we called "PEPM", Diethylphosphatoethyl methacrylate, had the best combination of fire retardancy, high grafting conversions, stability to laundering, etc.

PEPM, at about 35% add-on levels on 100% cotton, gives about 3% phosphorus content for fire retardancy and appears to enhance the laundering abrasion resistance. Since it was not commercially available, we synthesized it by making DECP (Diethylchlorophosphate) and reacting this with HEMA (Hydroxyethyl methacrylate) to make PEPM.

Grafted Product Properties

Fire Retardancy We have known for several years that PEPM could be grafted on cotton substrates to impart a fire retardancy which seemed to be outstanding. At 3% phosphorus content or 35% add-on (26% polymer content) with PEPM, even light fabrics are self extinguishing. Previously, photographs were shown (1) of adjacent loose fiber balls, grafted and ungrafted, ignited from a single source. The PEPM graft was self extinguishing next to the consumed control. DOC FF 3-71 char lengths are a function of fabric weights and constructions, so little purpose is served to quote char lengths, or show pictures to illustrate 3 to 6 inch chars at 3% phosphorus levels.

Since the phosphorus compound is grafted to the fiber, it is not surprising that the phosphorus assay remains relatively constant through 50 laundering cycles.

Abrasion Resistance The presence of nitrogen, from either a comonomer or from an over-padded resin, because of its synergistic action in the presence of phosphorus, permits lower levels of phosphorus to be used. However, acrylamide tends to reduce the normally good abrasion resistance of cotton when applied in resin formulations, or when grafted on the cotton. It had been observed that PEPM appeared to improve abrasion resistance in other studies. It was also known that a minimum of 18% add-on of PEPM (1.6% phosphorus), plus 13.3% add-on of acrylamide (2.0% nitrogen) would render cotton fabric flame retardant. Experiments were therefore set up to measure laundering abrasion weight losses on controls and grafts with PEPM-acrylamide copolymers, both with and without a durable press resin topping. Cotton muslin, 2.27 oz./yd^2, and cotton flannelette, 3.28 oz./yd^2, were grafted with PEPM-acrylamide comonomers to levels well in excess of the fire retardant minimums in order to exaggerate the effects on laundering. Portions of both grafts and controls were topped with PERMAFRESH 183 (Sun Chemical Corp.) durable press resin, prior to the laundering abrasion studies. Laundering was under home laundering conditions of 19 minutes wash at 45-50°C., using household detergent, 150-200 ppm average water hardness, 40°C.

rinses, tumble dried.

Although all the control samples burned their entire lengths, including those that were resin topped, all the grafts passed 50 HLTD with conditioned (not bone dry) char lengths of less than four inches.

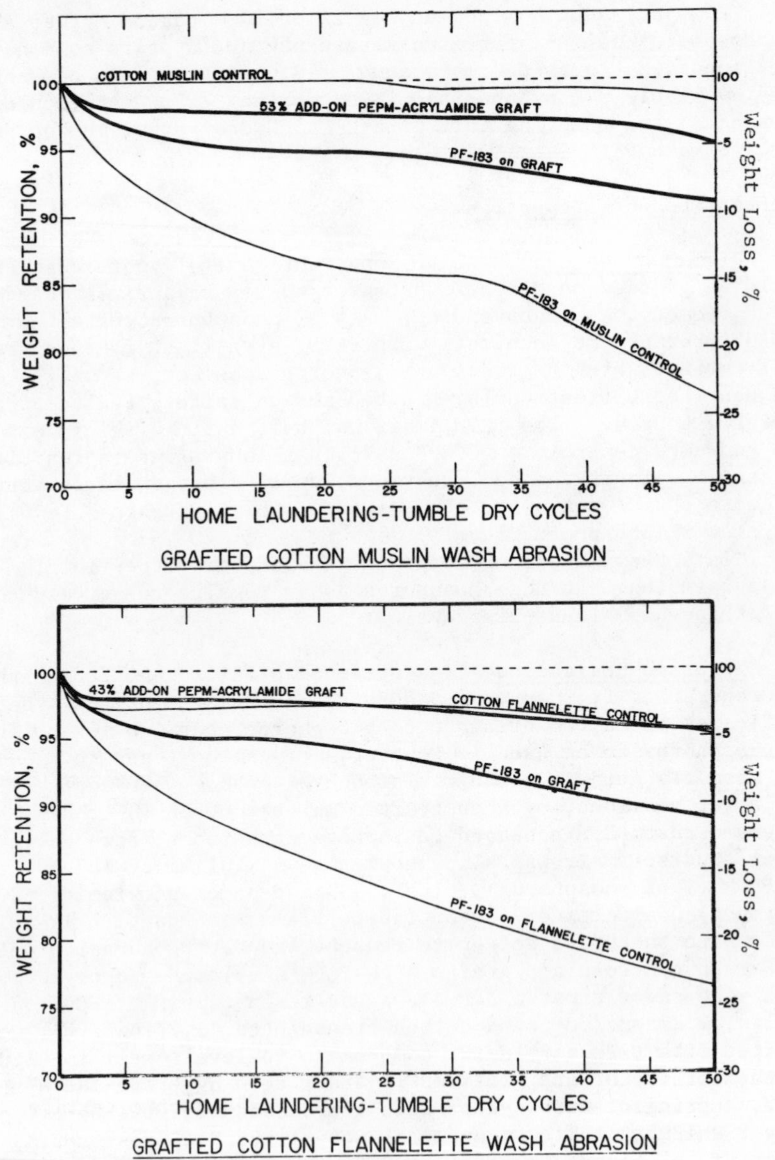

GRAFTED COTTON MUSLIN WASH ABRASION

GRAFTED COTTON FLANNELETTE WASH ABRASION

The grafted muslin was only 65% cotton and picked up 7.7% resin (on cellulose basis), while the control picked up 8.8% resin on its cellulose content. The cotton flannelette graft contained only 70% cotton and picked up 7.2% resin (on cellulose basis), while the control picked up 8.9% resin on the same basis. The decrease in weight losses due to abrasion after resin topping amounts to 12 to 14% as a result of the PEPM graft.

Rot Resistance Although the monomers were not selected for their rot resistance properties, soil burial trials were carried out on grafts employing PEPM and acrylic acid.

RESISTANCE TO SOIL ROT

	PERCENT STRENGTH RETENTION		
	INITIALLY	30 DAYS	90 DAYS
COTTON FLANNELETTE CONTROL	100	57	0
37·2% PEPM Add-on	101	84	58
33·4% ACRYLIC ACID Add-on	112	122	85
RAYON WOVEN CONTROL	100	31	0
21·6% ACRYLIC ACID Add-on	99	63	45

These trials show a high percentage strength retention of grafted cotton and rayon fabrics after extended periods of burial in moist soil, under conditions which caused the controls to fail completely in three months. Other monomer systems behaved similarly (2).

Disperse Dye Affinity Cotton does not normally dye with Disperse Class Dyestuffs. However, with PEPM grafts, polyester types of disperse dyestuffs are readily picked up, thereby presenting the possibility of eliminating the need for union dyeing and also avoiding the unsightly wear problems in blends of cellulosics with synthetics. Grafting on other functional groups may be used to attach other classes of dyestuffs, or to impart chemical resistance (2).

Dispersibility and Ion Exchange The ion exchange and cation scavenging abilities of carboxylic acids grafted on cotton fabrics were reported earlier (1). During the studies on the dispersibility of rayon fibers, it has been observed that calcium and magnesium ions associated with hard waters are rapidly removed by grafted fibers (3). These hard water ions will also reduce the dispersibility of the grafted fibers. When good soft water is used, sufficient carboxylic acid can be grafted onto textile length cellulosic fibers to permit excellent dispersion to be made in water, either in combination with, or without added wood pulp fibers, to produce non-woven sheets of good uniformity (1,3).

Summary

It should be kept in mind that grafting always adds permanent weight to a fabric. Thus, it is possible to increase weight, cover factor, fullness of hand, etc., while simultaneously imparting other desirable properties such as fire retardancy, abrasion resistance, etc., without loss of the original fabric strength.

SUMMARY

1. CONTINUOUS GRAFTING PROCESS - By Thiocarbonate Redox Methods

2. MICROWAVE ENERGY - Permits short reaction times

3. DEAERATED, SATURATED STEAM - Preferred with Microwave

4. WHEN ONLY STEAM IS USED:
   (a) Thin substrates are required
   (b) Surfaces remain ungrafted
   (c) Lower monomer conversions occur
   (d) Longer reaction times are required

5. PEPM MONOMER FIRE RETARDANT:
   (a) at 35% add-on for 100% Cotton
   (b) improved abrasion resistance in FR copolymers and DP treatments
   (c) makes Cellulose dye with Disperse Dyes

6. TEXTILE LENGTH FIBERS WATER DISPERSIBLE
   (a) Textile Fibers dispersed in water
   (b) Cations removed from solution by Ion Exchange

The basic procedure requirements for carrying out the thiocarbonate-redox grafting process as a continuous reaction have been summarized. Extremely short graft copolymerization reaction times have been made possible through the use of microwave energy to kick off the reactions. Our experience has been that resonant cavity type applicators can be used most satisfactorily, particularly in our double belt reactor units when deaerated, saturated steam is also present to keep the belts at reaction temperatures. When microwave is not used, it is still possible to graft using deaerated, saturated steam, but we have certain limitations. These are:

(a) Thick substrates do not graft uniformly, so we use only light weight fabrics.
(b) Surface grafting appears to be "killed" and grafting is almost completely absent from the surface fibers.
(c) Monomer conversions are usually lower than with microwave.
(d) Slightly longer minimum reaction times are required than is the case when microwave is used.

Diethylphosphatoethyl methacrylate, or PEPM, has been revealed for the first time as our superior fire retardant which, when grafted on cellulosics, makes them self-extinguishing at ca. 35%

add-on.

The abrasion resistance of cotton is not harmed by the presence of PEPM grafts and, in the case where a resin after-treatment is used, the resistance to laundering abrasion is greatly improved. Resistance to attack by soil bacteria and new dyeing capabilities are supplementary benefits.

Textile length rayon fibers have been made water dispersible through grafting with monomers which contain carboxylic acid groups (3,15). These are capable of simultaneously acting as cation scavengers. The ion exchange capacity can be varied at will, depending on monomer and add-on level selected.

In our discussions of this continuous thiocarbonate-redox grafting method, we have again "revealed the tip of an iceberg"! Since it is such a flexible tool, the potential scope for the thiocarbonate-redox grafting is not limited to the few uses which I have mentioned. Instead, the only limits for this versatile chemical grafting process will be the length and breadth of vision exhibited by future investigators.

Acknowledgements

I would like to express my gratitude to Dr. R.W. Faessinger and all our other colleagues in the Organic Synthesis Group for their contributions which helped make this paper possible. I would also like to thank **Cotton Incorporated** and **Scott Paper Company** for permission to publish these researches.

Literature Cited

1. - Brickman, W.J., Tappi (1973) 56, (3), 97-100.
2. - Brickman, W.J. and Faessinger, R.W., J. Textile Chemists & Colorists (1973) 5, (5), 38-41.
3. - Teichmann, H.E., Brickman, W.J., Faessinger, R.W., Mayer, C.G. and Kraessig, H.A., Tappi (1974) 57, (7), 61-64.
4. - Belgian Patent No. 646,284, granted October 8, 1964.
5. - Belgian Patent No. 646,285, granted October 8, 1964.
6. - Belgian Patent No. 676,419, granted August 16, 1966.
7. - British Patent No. 1,059,287, issued February 22, 1967.
8. - Faessinger, R.W. and Conte, J.S., U.S. Patent No. 3,330,287, issued July 11, 1967.
9. - Faessinger, R.W. and Conte, J.S., U.S. Patent No. 3,340,326, issued September 5, 1967.
10. - Faessinger, R.W. and Conte, J.S., U.S. Patent No. 3,357,933, issued December 12, 1967.
11. - Faessinger, R.W. and Conte, J.S., U.S. Patent No. 3,359,222, issued December 19, 1967.
12. - Faessinger, R.W. and Conte, J.S., U.S. Patent No. 3,359,224, issued December 19, 1967.
13. - Kraessig, H.A., Das Papier (1970) 24, (12), 926-934.

14. - Kraessig, H.A., Svensk Papperstidning (1971) 74, (15) 417-428.
15. - Gotschy, F. and Kraessig, H.A., Das Papier (1972) 26, (12), 813-817.
16. - Dimov, K. and Pavlov, P., J. Polymer Sci. (1969) A-1 7, 2775-2792.
17. - Samojlov, V.I., Morin, B.P. and Rogovin, Z.A., Faserforschung und Textiltechnik (1971) 22, (6), 297-304.
18. - Morin, B.P., Vilkov, V.A. and Rogovin, Z.A., Faserforschung und Textiltechnik (1973) 24, (2), 77-82.
19. - Tjuganova, M.A., Lischevaskja, M.O. and Rogovin, Z.A., Paper presented at 12th International conference, Dornbirn, Austria, September 18-20, 1973.
20. - Rogovin, Z.A., Tappi (1974) 57, (7), 65-68.
21. - Kokta, B.V. and Valade, J.L., Tappi (1972) 55, (3), 366-369.
22. - LePoutre, P., Hui, S.H. and Robertson, A.A., J. Applied Polymer Sci. (1973) 17, 3143-3156.
23. - Ehrnrooth, E., J. Polymer Sci. (1973) Symposium No. 42, 1569-1582.
24. - Dutch Patent No. NL 7,201,385, granted August 4, 1972.
25. - West German DT-OLS No. 2,204,442, issued August 10, 1972.
26. - "Theory and Applications of Microwave Power in Industry: A Short Course", published by the International Microwave Power Institute, P.O. Box 1556, Edmonton, Alberta, Canada, 1971.
27. - Gerling Moore, Inc. (formerly Genesys Systems, Inc.), 1054 East Meadow Circle, Palo Alto, CA 94303 (catalogs and Instruction Manual for Model 4003 Variable Power Source).

3

Cellulose Derivatives; Polymers with a Future

CLAYTON D. CALLIHAN

Louisiana State University, Baton Rouge, La. 70802

Summary

The shortage of oil and natural gas will soon be reflected in shortages and spiraling prices for polymers based on hydrocarbons. Cellulose derivatives, which will be their replacement, are currently not competitive in many applications. This paper presents data on the fundamental relationships between several derivatives and their solubility characteristics. It also discuss' the design of reactors needed to produce the polymers most economically.

Introduction

M. King Hubbert, Chairman of the U.S. Energy Resources Study, reported the data collected by he and others in the book, Energy Resources, published in 1962. Figure 1 shows the gloomy picture predicted by their research. Notice that the actual rate of crude oil production follows a jagged line on the left side and this line is dashed for the predicted production. The bottom line shows that the rate of increase of proven reserves started to fall shortly before 1950 and is still falling. These 1961 data have been upgraded by Dr. Hubbert and his group and shows that the shape of the curves still hold very well, however, the peaks have been shifted up and slightly to the right. A shift caused by the excitement and lack of preparedness for an oil shortage which lead to drastic increases in exploration.

Figure 2 taken from the same report shows the United States Production of Natural Gas with both high and low estimates of ultimate reserves. These data have also been updated with some more recent numbers, but the ultimate production appears to approximate the upper curve. The important thing to stress here is not the absolute values of these curves nor even the timing of the peaks, but the inevitable demise of production of a non renewable resource.

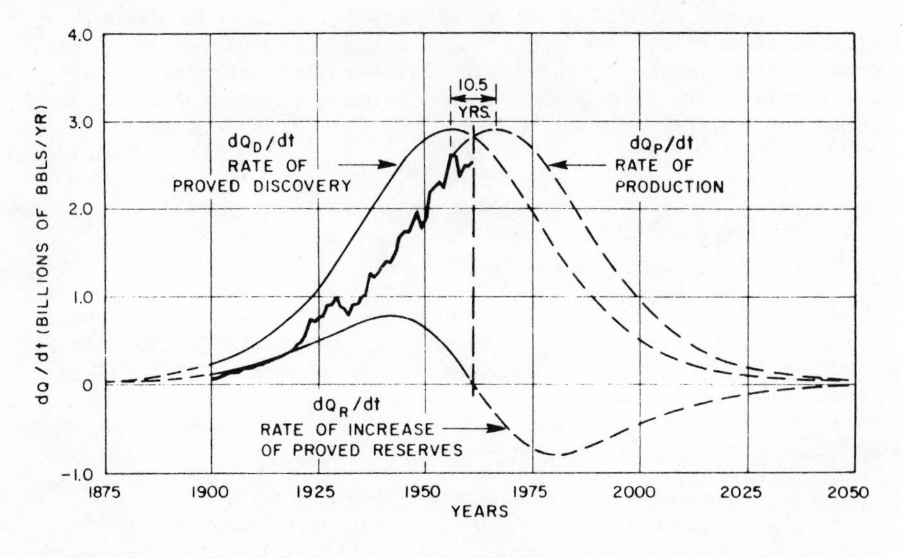

Energy Resources

Figure 1. Rates of discovery, production, and increase of proved reserves of U. S. crude oil (15)

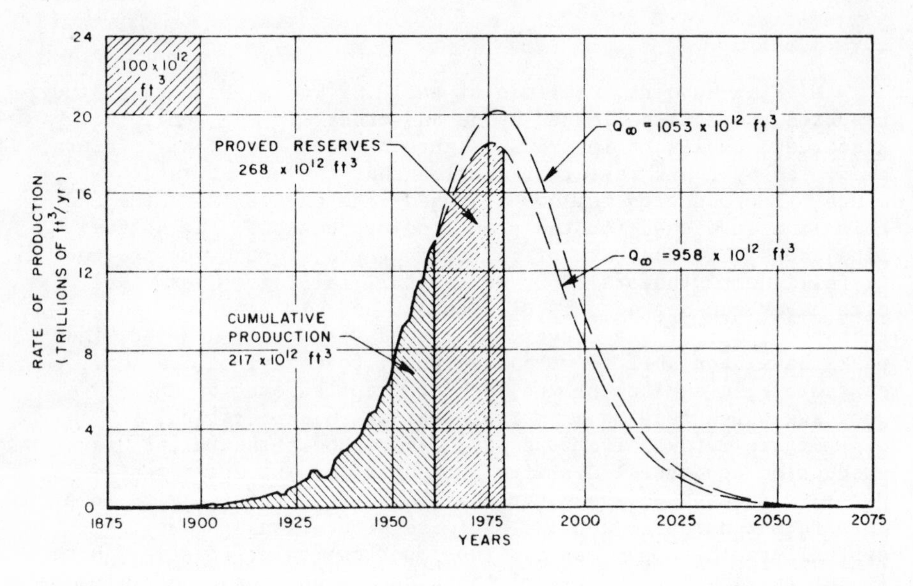

Energy Resources

Figure 2. U. S. production of natural gas for high and low estimates of ultimate reserves (15)

The total time to which a fuel may be exploited to some trivial amount is not a significant figure when placed in a reasonable time perspective. Figure 3 shows that exploitation of the fossil fuels from beginning to ultimate exhaustion will comprise but a brief episode. These data further show that increases in price may lead to a slight redistribution in where the world's oil supplies will be consumed but will have no effect whatever in changing the world's total available supply.

With the demise of fossil fuels another energy source will probably be found to take its place such as atomic energy, solar energy, or whatever, however, there is no possibility of finding organic deposits that can serve as a new basis for hydrocarbons. One can expect then that in a relatively few years, polyethylene, polypropylene, styrene, butadiene etc. are going to either disappear or be so expensive that they will find little application in our daily lives. The obvious replacement for the organic base of these polymers is cellulose because of its perpetual nature.

The technology for the conversion of cellulose to plastics, to fibers, to water soluble derivatives, to films, to lacquers and paint bases, etc. is already reasonable well known. The lack of universally large scale production of many of these cellulose based polymers is due to their relatively high cost. Still they fall only a little short of competing economically in many fields and, of course, do compete in a great many others. Only small improvements in manufacturing costs and product quality would make the cellulose based polymers competitive in a great many more applications. The impending unavoidable price increases in hydrocarbon based polymers will mean that cellulose based polymers will soon compete in many new areas without any improvements in manufacturing procedures. To forstall a dramatic decrease in our polymer consumption, however, researchers should turn their attention to cellulose and to its use as replacement polymers for those that are now derived from oil and gas.

Organic considerations. Based on its structure, cellulose should be water soluble but strong hydrogen bonding prevents solvation by water except at extremely high temperature. Theoretically, at high temperatures the cellulose molecules possess so much kinetic energy that they shake free from the orientation required for good bonding and become soluble in water.

Native cellulose, such as wood, cannot be easily substituted or converted to cellulose derivatives because the lignin associated with it must first be removed. This is done commercially by cooking the cellulose at high temperature with either sulfur dioxide or sulfur trioxide and caustic. The final traces of color are removed by a bleaching process which uses either chlorine dioxide or sodium hypochlorite to convert the remaining lignin to water soluble material that is subsequently removed by washing.

It is interesting to note that during this cooking process the cellulose crystallinity is completely destroyed and only reforms again on drying. The dried pulp is then shipped to chemical companies where the crystallinity must be redestroyed by strong reagents in order to uniformly substitute on it. Strangely enough, no one seems to take advantage of shipping the pulp with 15 to 20 percent moisture even though to do so would cut down greatly on the amount of reagents required to make homogeneous cellulose derivatives.

An example of the above oversight can be cited in the preparation of paint grade carboxy methyl cellulose. This derivative is water soluble at a DS of approximately 0.3, however, Chemical Grade CMC invariable has 3 or 4 times this substitution just to remove minute traces of unsolvated fibers. The use of "wet" cellulose would save greatly on the required DS to obtain complete fiber removal.

According to Klug (1), substituted cellulose loses its crystallinity at a very low degree of substitution (DS). For hydroxyethyl substitution this may be as little as 0.05 and for larger substituents this may even be less. The reason for this exceptional behavior can be found in the relative rigidity of the cellulose chain backbone. Figure 4 shows that if the backbone is rigid a substituent group may force many neighboring anhydroglucose units away from their positions of maximum intermolecular bonding.

Structural Variables. The physical properties of substituted cellulose or cellulose derivatives depend upon at least the following factors:

A. Molecular weight and molecular weight distribution

1. oligomers have no strength but serve as plasticizers during fabrication.
2. extremely high molecular weight with narrow distribution is difficult to fabricate but strong if properly oriented

B. Regularity of substitution on chains

1. irregular substitution allows chains to hydrogen bond over short segments
2. regular substitution leads to the desired properties of the derivative with the least reagent cost

C. The Bulk or size of the substituent

1. the larger the size of the substituent group the further the chains are separated and the less hydrogen bonding will occur

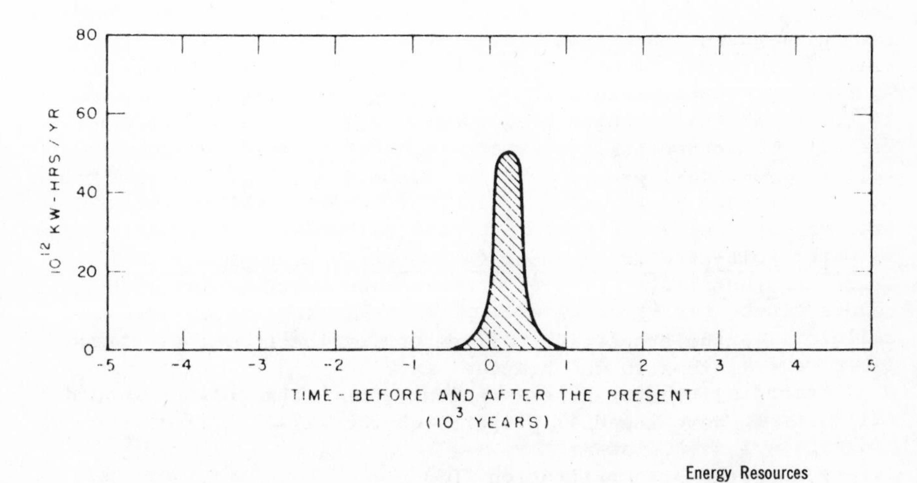

Figure 3. Total world production of fossil fuels in time perspective (15)

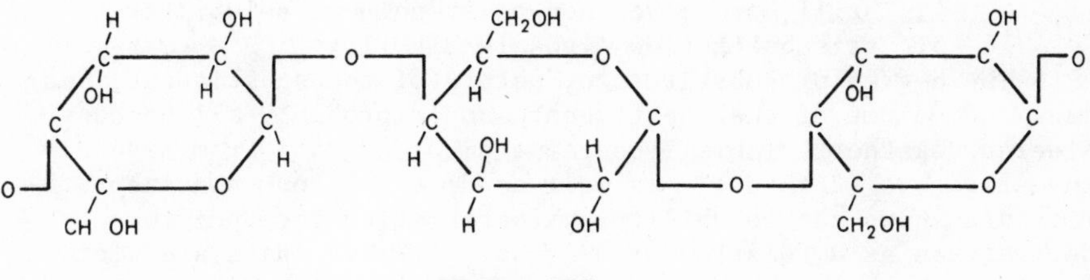

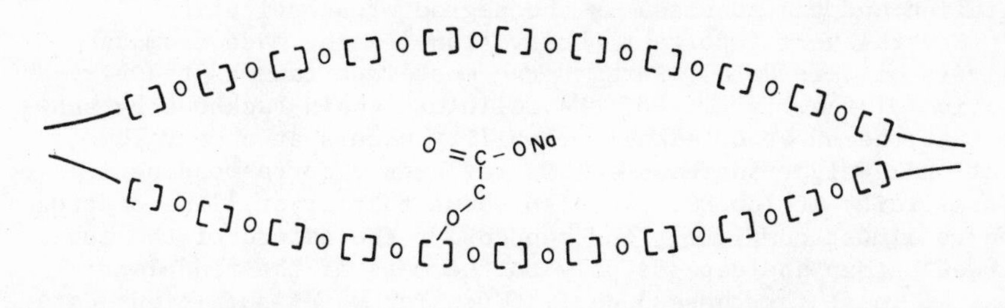

Figure 4. Decrystallization caused by protruding chain branches on cellulose molecules

2. the smaller the size the more substitution necessary to get the desired effect

D. <u>The nature of the substituent</u>

1. substituents containing hydrogen donors or acceptors lead to higher DS requirements to obtain the same objective
2. alkyl groups lead to solubility in non polar solvents at lower DS

E. <u>Multiplicity of substitution on the same hydroxyl group</u> (MS)

1. the properties will depend on the number of carbon atoms in the branches
2. if the branches are long enough the polymer should have the dual property characteristics of graphs

F. <u>Degree of substitution</u> (DS)

All cellulose derivatives go through the following stages as the DS in increased:
1. low substitution gives alkali solubility
2. slightly more leads to water solubility
3. more substitution gives polar solvent solubility
4. still more gives non-polar solvent solubility
5. trisubstitution gives insolubility

The degree of substitution, nature of the substituent, and mass and volume of the substituent should probably all be considered together. Table I contains smoothed data from several investigators (2 thru 13) as well as data from our own laboratories and shows the solubility characteristics for 3 different derivatives as a function of DS. Some of these data are plotted in Figure 5 where the molar ratio of carbon to oxygen in the final derivative is plotted as a function of DS for the various solubility ranges. Data from the same table are plotted in Figure 6 but here the ordinate is the relative mass after substitution and the abscissa is the degree of substitution.

Several very important observations can be made from an analysis of this data. Perhaps the most important is the verification of the rigidity of the cellulose chain backbone by showing that the unset of alkali solubility occurs at a very low fractional weight increase. This reflects a corresponding loss of crystallinity at low DS. It also shows that crystalline destruction is almost completely independent of the nature of the substituent group and depends only on the mass of the pendulant group at least for those 3 derivatives for which sufficient data are available to tell. A further rather unexpected observation is that the unset of water solubility is also almost an exclusive

Table I

Solubility Characteristics of Various Cellulose Derivatives

	(14) Methoxyl OCH_3			(28) Ethoxyl OC_2H_5			(80) Carboxy Methyl OCH_2COON_a		
	DS Range	Carbon To Oxygen Ratio (Molar)	Relative Mass After Substitution	DS Range	Carbon To Oxygen Ratio (Molar)	Relative Mass After Substitution	DS Range	Carbon To Oxygen Ratio (Molar)	Relative Mass After Substitution
Insoluble except in special solvents like DMSO & NO_2 etc.	0-0.3	1.2 -1.26	1.0 -1.03	0-0.17	1.2 -1.27	1.0 -1.03	0 - .05	1.2 -1.2	1.0 -1.03
Soluble in 4 to 8% NaOH	.4-1.2	1.28 -1.44	1.04 -1.10	.25- .70	1.30-1.48	1.04-1.12	.10 - .25	1.19-1.18	1.05-1.12
Soluble in Cold Water	1.3-2.2	1.46 -1.64	1.11 -1.19	.80-1.30	1.52-1.72	1.14-1.22	.30 -1.40	1.18-1.13	1.15-1.69
Soluble in polar organic solvents	2.1-2.6	1.62 -1.72	1.18 -1.22	1.40-1.80	1.76-1.92	1.24-1.31	2.2 -2.8	1.11-1.09	2.08-2.38

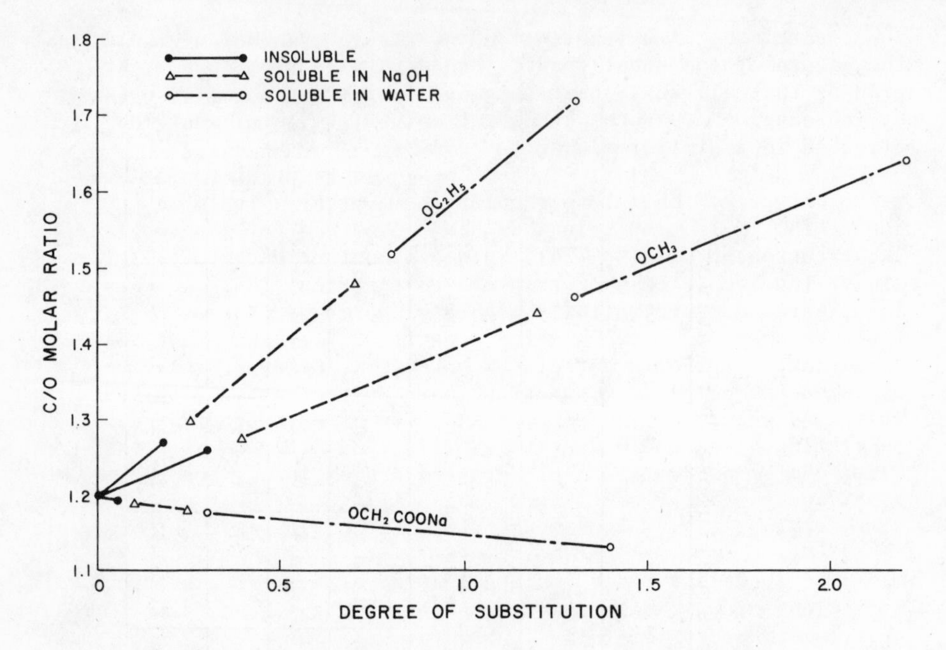

Figure 5. Solubility characteristics of various cellulose derivatives

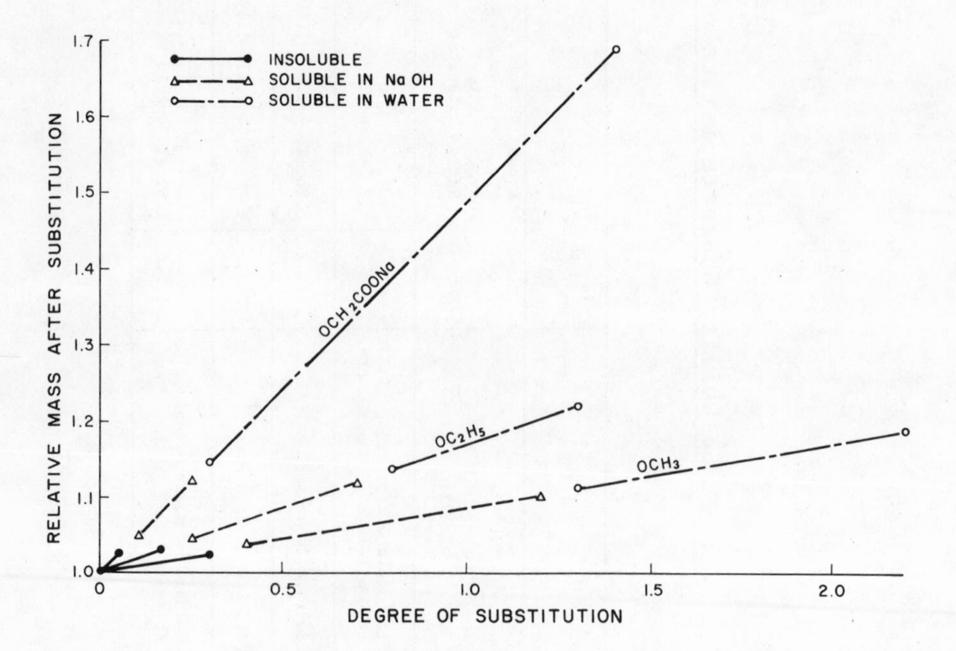

Figure 6. Solubility characteristics of various cellulose derivatives

function of the mass increase and nearly independent of the DS or the nature of the substituent. Since this is true, one might predict that all subsequent responses which are normally obtained by increases in DS such as solvent solubility, etc. would be affected in a similar manner.

The data of Table I might be extended to predict, for example, that propyl cellulose should become water soluble at a DS of about 0.5. This prediction has been confirmed for isopropyl substitution by Calkins (14). Since substitution onto cellulose always involves a loss of reagents to side reactions and this loss increases exponentially with DS, one might predict that it would be cheaper to use a large substituting group rather than a small one. It should be pointed out again, however, that low DS leads to non-uniform substitution. Furthermore, if the substituting group were an alkyl chain, its effectiveness as a chain separating agent would decrease. This is due to the fact that alkyl chains have freedom of rotation around the carbon to carbon bonds and therefore sections of the chain could move out of the path to allow closer approach of the neighboring cellulose backbones.

The above analogies would not hold for hydroxy alkyl substitutions because here a new site for hydrogen bonding is added everytime a hydroxyl group disappears. The table would predict, however, that hydroxy ethyl substitution would require the largest DS to obtain water solubility of any substituting group and this is indeed true.

If high molecular weight derivatives are to be produced the starting material is usually purified cotton linters. If low molecular weight is desired then wood pulp is generally used in place of cotton. The reason for this is not that the linters cannot be depolymerized to the desired molecular weight but because wood pulp is cheaper and less crystalline.

In the older type plants the molecular weight of ethers was controlled by first dipping the cellulose into caustic solution and then subjecting this material to elevated temperatures. It is interesting to note that alkaline cellulose cannot be hydrolyzed in the presence of heat alone but also requires oxygen. This fact allows the molecular weight of the cellulose to be controlled by admitting measured quantities of air into the reactor while the derivative is being made. In the case of cellulose esters, the molecular weight is controlled by the amount of water present during the reaction.

The symmetry or regularity of substitution of the hydroxyl groups in etherification of the cellulose depends on several factors such as the uniformity of alkali distribution, the reagent concentration, the reagent volume, etc. Very rapid reactions give extremely non-uniform substitution. Freshly prepared alkali cellulose also gives unsymmetrical substitution. Since non-homogeneity of substitution implies that certain sections of the cellulose chains will contain no substituent groups at low substitution, this means that higher substitution will be required to eliminate unreacted fibers and to give uniform physical properties.

Reactor Design. The substitution on the hydroxyls of cellulose with either ester or ether groups causes the release of large amounts of heat. Since most of it is liberated early in the reaction cycle when reactant concentrations are high the early control of temperature and therefore pressure is difficult. In almost every case, an effort is made to maintain a more or less uniform rate of reaction by taking advantage of their temperature dependence. The rate of substitution for most reactants follow quite closely an Arrhenius relationship and normal second order kinetics. Initially, when the reactant concentrations are very high, it is extremely difficult to remove enough heat to hold the temperature down; later on in the reaction cycle, it becomes difficult to increase the temperature sufficiently without exceeding pressure limitations to force the reaction to completion.

Because the overall heat transfer coefficient between the solid cellulose inside the reactor and the water flowing through the shell is very low, most older cellulose reactors are rather uneconomically designed to compensate for the danger inherent in this situation. A typical example is in one of the commercial production units for ethyl cellulose where a carrier solvent of benzene and alcohol is employed to act as a heat transfer agent for the solid cellulose. Here, the heat transfer coefficients are very good but the rate of reaction is extremely poor, especially during the latter stages of reaction when an attempt is being made to push it to completion. The dilution effect of the solvent on the reactant concentrations extends the reaction time several hours. The use of a carrier solvent also causes several subsequent unnecessary distillations and separations with the inevitable loss of considerable benzene and alcohol.

Many different reactor designs have evolved throughout the years some, seemingly, without serious thought given to fundamental engineering analysis. When it became apparent that removal of the carrier solvent was necessary for cost reductions, it still left a serious problem of how to get the heat transferred to the water in the shell without causing a runaway exotherm. The most obvious answer was to increase the heat transfer area by putting additional tubes inside the reactors. This practice uses up expensive reactor volume and usually causes the formation of "hot spots" due to poor mixing.

Perhaps the least likely of the second generation designs was the spherical-rotating-solid-phase reactor. A sphere by definition offers the least heat exchange area per unit volume of reactants of any design. Its only redeeming feature is its ability to withstand pressure. Obviously, a long thin tube would provide both a resistance to rupture by pressure and a large amount of heat transfer area per unit volume of reactants.

Unfortunately, the requirement for product homogeneity requires good end to end mixing and so the length to diameter ratio must be set by this requirement. Said another way; the reactor must give good end-to-end mixing.

The third generation reactors have made much better use of engineering fundamentals. They do not use carrier solvents for reasons already discussed. They use instead an excess of one of the liquid reactants thereby making certain the reaction can be forced to completion at the proper time. To aid the end-to-end mixing, a helical ribbon is used inside the reactor. This ribbon also scrapes the internal heat transfer surface thereby assuring maximum transfer coefficients. Another development of the third generation reactors is the use of an external heat exchanger. These units provide external cooling and condensation with maximum heat transfer coefficients and minimum costs for transfer area since it consists merely of a standard shell and tube bundle.

The reaction time required for producing a given derivative is primarily dependent on the rate of heat transfer designed into the reactor. The cost of a typical cellulose ether reactor is currently in the neighborhood of a million dollars because all of the metal in contact with the alkali must be either nickel or monel.

At this cost, it is highly desirable to shorten the reaction time as much as possible. The external tube bundle mentioned previously, and used on the third generation reactors need not be an expensive alloy since only liquids are vaporized. This means therefore, that by the use of a much larger tube bundle then would normally be required, the reaction time can be cut by a factor of 2 or 3.

Reaction Efficiency. The efficiency of substitution is extremely low in the manufacture of most cellulose derivatives. Usually more than twice the theoretical amount of reagents are required to produce a given derivative. Here again, part of the problem is a result of poor engineering practices. As an example of this, when preparing ethyl cellulose, one produces as a by-product, first ethyl alcohol, and then ethyl ether. Obviously, the ether could not form without the presence of alcohol, furthermore, the formation of ether occurs at a fast rate only when the concentration of caustic is low. The caustic is diluted both by water formed as a product of the etherification reaction and by water released when the caustic is converted to salt.

Removal of this water is possible by inserting several plates between the overhead condenser and the reactor to remove part of the water by fractionation. Here the heat of reaction supplies part of the heat needed for reflux while heat can also be supplied through the jacket of the reactor if desired. An exact analogy can be drawn from the manufacture of methyl cellulose. The reaction starts very rapidly at 42°C and is difficult to control until the caustic concentration is cut to roughly one half its original value. During this period considerable methyl alcohol is taken to the overhead condenser by using the heat of reaction. After that, the reaction is forced by gradually increasing the jacket temperature to about 100°C, and reflux during this period is provided by external heat.

Conclusion

At one time expenditures for research on cellulose derivatives represented a considerable fraction of the total plastics research budget for several large companies. This work has been gradually phased out or greatly scaled down because the rewards were seen to be greater for research on new polymers based on hydrocarbons. The impending shortage of oil and natural gas will soon reverse this trend and start a resurgence in cellulose research. A great variety of products with a tremendous range of physical properties is already possible from cellulose but even these are only a beginning. Exotic new derivatives have been made in our laboratories and in the laboratories of other investigators. Research will bring lower prices and better performance for the only polymer we will be able to count on in the future.

Literature Cited

1. Klug, E. D., Starke, (1961), 13, 429.
2. Gladstone, J. H., J. Chem. Soc. (1852), 5, 17.
3. Champetier, G., Ann Chim, (1933), 20, 5.
4. Sobue, H., et. al., Z. Physik. Chem., (1939), B43, 309.
5. Clark, G. L., Ind. Eng. Chem., (1930), 22, 474.
6. Vieweg, W., Ber., (1907), 40, 3876.
7. Shutt, R. S., U.S. Patent 2,37,359, (Mar. 13, 1945).
8. Hess, K., et. al., Z. Physik. Chem., (1931), B14, 387.
9. Lieser, T., et. al., Ann., (1936), 56, 522.
10. Brownsett, T., J. Textile Inst., (1941), 32, T32, T57.
11. Lovell, E. L., Ind. Eng. Chem., Anal. Ed., (1944), 16, 683.
12. Bock, L. H., et. al., U.S. Pat. 2, 083, 554, (June 15, 1937).
13. Mahoney, J. F., et. al., J. Am. Chem. Soc., (1942), 64, 9.
14. Calkins, J. B., J. Phys. Chem., (1936), 40, 27.
15. Hubbert, M. K., Energy Resources, (1962).

4

Silyl Cellulose

ROBERT E. HARMON and KALYAN K. DE

Western Michigan University, Kalamazoo, Mich. 49001

Early Works on Silylation of Cellulose

The silylation (the term ''silylation'' is used in the general sense of substitution with triorganosilyl groups) of numerous natural products, including various sugars, has been reported in the literature, and this procedure has become a commonplace operation whenever the effects of hydrogen bonding on the physical properties are undesirable.

One of the most ubiquitous and important natural products with an abundance of free hydroxyl groups which significantly determine its physical properties is cellulose. Replacement of some or all of the hydroxyl protons of cellulose by silyl groups can be expected to alter radically the properties of this polymer, just as esterification or alkyl ether formation drastically modify the parent cellulose. The effects of polymer modification by silylation have been demonstrated in at least two cases, eg., silyl polyvinyl alcohol (1) and silyl polyureas (2). The increased solubility of silylated polymers in nonpolar solvents is particularly noteworthy.

Silicon tetrachloride and aryl- and alkylhalosilanes react easily and rapidly with organic hydroxyl groups (3,4,5) to yield hydrogen chloride and silicic esters. A number of patents have been issued dealing with the treatment of cellulose with organosilicon halides to impart water repellency (6,7,8). In one case (6) a surface reaction either with adsorbed moisture or with hydroxyl groups of the cellulose was postulated. However, the amount of product formed was too small to be measured or analyzed. Jullander (9) has studied the reaction of silicon tetrachloride with nitrocellulose and reports the formation of gels due to cross linkage.

Schuyten (10,11) and Hunter (12) have reported the preparation and modified properties of partially silylated cellulose. The ether or ester groups in partially etherified or esterified cellulose are partially replaced with $R_2R'Si$ radicals (12), where R is aryl and R' is aryl or alkyl, to give film-forming cellulose derivatives with improved thermal stability, dielectric strength, and moisture resistance. The silicon content of this product reported by Hunter was 5.44%. Schuyten has obtained trimethylsilyl cellulose with as high as 2.75 trimethylsilyl groups per glucose unit. With partially substituted cellulose acetate, the total substitution approached 3.0 (Table I).

Table I. Reactions of Alkylchlorosilanes With Cellulose Acetate

Reagent	Acetyl groups/glucose		Silyl groups/glucose	Total groups
	Before	After		
$(CH_3)_3SiCl$	2.30	2.24	0.65	2.89
$(C_2H_5)_3SiCl$	2.30	2.24	0.58	2.82
$(CH_3)_3SiCl$	2.90	2.87	0.14	3.01

Various other silanes were employed and similar results were obtained (Table II).

Table II. Reactions of Various Substituted Silanes With Cellulose

Reagent	Cellulose (g)	Si (%)	Silyl group/glucose
$(C_2H_5)_3SiCl$	2.15	11.55	1.27
$(CH_3)_2SiCl_2$	1.79	2.47	a
$n\text{-}C_8H_{17}SiCl_3$	1.96	1.33	a
$CH_3Si(OAc)_3$	1.90	1.16	a

[a]Not calculated because of the possibility of cross linkage.

The trimethylsilyl cellulose obtained might be decomposed by boiling with water or with dilute acid or base. No quantitative results were obtained but boiling with dilute acid yielded a volatile liquid product which burned and deposited silica and was assumed to be trimethylsilanol. Accordingly samples of trimethylsilyl cellulose containing 2.6 trimethylsilyl groups per glucose unit were placed in open containers under three sets of conditions: (1) desiccated over phosphorus pentoxide; (2) dried in an oven at 105°; and (3) conditioned at 70°F (21.1°) and 65% relative humidity. The samples were weighed at intervals and the percent retention of added weight was plotted against time. The results are shown in Figure 1. The inflection at the beginning of the curve for the conditioned sample is due to increase in moisture content at 65% relative humidity.

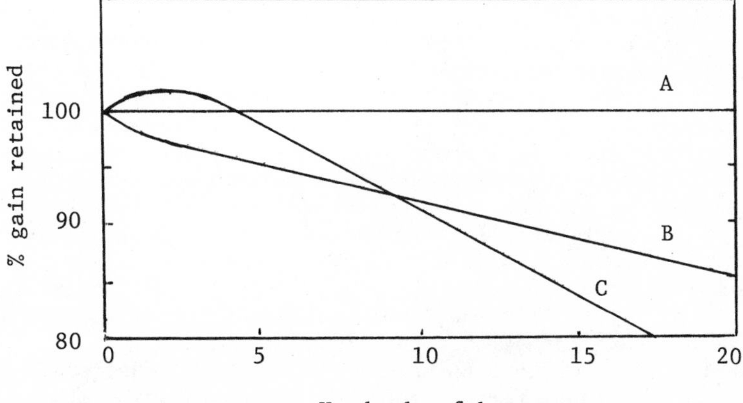

Figure 1. Deterioration of trimethylsilyl cellulose: A, desiccated (P_2O_5); B, heated (105°); C, conditioned at 70°F and 65% relative humidity

These data indicate the formation of definite compounds of cellulose with substituted silanes.

A 0.1 g. sample of trimethylsilyl cellulose (2.42 groups per glucose unit) was placed in 5 ml. of solvent in a small tube and tumbled on a wheel for 24 hours at room temperature. When the major portion of the material remained undissolved as determined by visual inspection, it was considered insoluble. In some cases (designated by ''sw'') the sample particles showed a tendency to swell. No noticeable amount dissolved in the following solvents: acetone, benzene, carbon tetrachloride, chloroform, diethylformamide, ethyl acetate, ethyl alcohol, ethylene dichloride, ether (sw), methylene chloride, methyl ethyl ketone (sw), nitromethane, nitropropane, pyridine (sw), s-tetrachloroethane and xylene. The material was not soluble in the following mixtures: acetone (80%), ethanol (15%), ethyl acetate (5%) (sw); toluene (80%), ethanol (20%) (sw); ethylene dichloride (90%), methyl alcohol (10%); and benzene (66%), ethylene dichloride (34%) (sw).

Rogovin (13,14) has reported preparing silyl cellulose from alkali cellulose (1 part sulfite pulp + 10 parts 40% NaOH, pressed to 4 times the weight of cellulose) and halomethyltrialkylsilanes. Maximum degrees of substitution were reached when the reaction was carried out at 120° for 20 hours in sealed ampules. Reaction with chloromethyltrimethylsilane under these conditions gave silyl cellulose with 10.9% silicon, with iodomethyltrimethylsilane gave silyl cellulose with 13.9% silicon, and with chloromethyltriethylsilane gave silyl cellulose with 4.5% silicon. The products were insoluble in common organic solvents and in cuprammonium solution.

Silylation With Bis(trimethylsilyl)-Acetamide (BSA)

Klebe and coworkers (15-18) have prepared a number of trialkylsilyl derivatives of cellulose by using silyl amides. The favorable results obtained with bis(trimethylsilyl)-acetamide (BSA) as a silylating agent for various classes of compounds with reactive protons (19) prompted them to attempt the silylation of cellulose with this reagent and to reinvestigate the properties of the product. The silylation of cellulose with BSA turned out to be straight forward once a suitable solvent was found. It is characteristic of highly hydrogen-bonded materials of high molecular weight that their silylation is very sluggish, even when the material in the silylated form is quite soluble in the solvent used for the reaction. Solvent systems in which the parent polymer is somewhat soluble or in which at least some swelling occurs are preferable even if the solubility of the silylated product is only marginal in the particular solvent.

Silylations have been carried out by suspending various grades of natural cellulose in polar solvents such as dimethylsulfoxide (DMSO), dimethylformamide (DMF), N-methylpyrrolidone (NMP), and hexamethylphosphoramide (HMPA), with addition of an excess of 20-30% over the calculated amount of BSA and on heating the agitated mixtures under anhydrous conditions at temperatures of 100° to 150°C. HMPA and NMP were found particularly useful. In the latter solvent the fibers of a wood pulp cellulose turned into a transparent tan-colored gel within one hour at a temperature of 150°C. This NMP-insoluble gel yielded a viscous solution upon addition of xylene or benzene to the mixture.

The choice of the grade of cellulose proved to be of some consequence. Depending on the prior history of the cellulose, the silylations go more or less readily to completion. In some cases, although most of the starting material appeared to have reacted, some insoluble gel remained. Some commercially available wood pulp celluloses (types V-60 and V-90, Buckeyee Cellulose Corp., Memphis, Tennessee) are very readily silylated. Cotton linter pulps from the same source were generally of higher molecular weight and, although they left no unreacted insoluble material, gave extremely viscous solutions.

The nature of the hydroxyl groups in cellulose suggests that any resistance to silylation may be due more to steric reasons and lack of solubility than to intrinsic "chemical" difficulties in displacing these particular protons by trimethylsilyl groups. Weaker silylating agents like silylamines could be expected to suffice once suitable reaction conditions were found. Treatment of cellulose in NMP with N-trimethylsilylpiperidine at 140-150°C for 3-5 hours gave viscous solutions with very little insoluble residue.

The polymeric product prepared with any of these silylating agents and solvents could be recovered either by vacuum distillation of volatile matter, or more conveniently, by precipitation with polar solvents like acetone, acetonitrile, or alcohols which yielded the polymer in the form of white fibers. The products contained 2-3 trimethylsilyl groups per repeating unit and were all cases soluble in aromatic, chlorinated, and a number of aliphatic solvents.

Silylation With Trimethylchlorosilane

The composition of the soluble trimethylsilyl cellulose obtained by Klebe and coworkers (15-18) must be virtually identical (according to the elemental analysis) with the totally insoluble product which Schuyten and coworkers (10,11) obtained by treatment of cellulose with trimethylchlorosilane in pyridine. Freshly distilled commercial trimethylchlorosilane and a sample of the cellulose which had yielded soluble product with BSA also gave an insoluble product. The cross-linking of the polymer therefore occurred as a result of either some secondary mode of reaction of trimethylchlorosilane itself or the presence of some impurity in the chlorosilane. The most likely impurities in this commercial product are higher chlorinated silanes. Dimethyldichlorosilane and methyltrichlorosilane boil at 70° and 66°C, respectively, and it should be very difficult to remove by fractionation traces of these compounds from trimethylchlorosilane with a boiling point of 57°C. Therefore, pure trimethylchlorosilane was prepared by reaction of anhydrous hydrogen chloride with hexamethyldisilazane and allowed to react with cellulose suspended in a mixture of pyridine and xylene. A completely soluble product was indeed obtained after a reaction time of 4 hours at 110°C. Thus, the insolubility of the product in the earlier experiments had evidently been caused by impurities in commercial trimethylchlorosilane.

The likely culprits in this undesirable cross-linking reaction, dimethyldichlorosilane and methyltrichlorosilane, are known to be considerably more reactive than trimethylchlorosilane. Addition of a small amount of sugar to the mixture of commercial trimethylchlorosilane and pyridine and heating for a few minutes prior to the addition of the cellulose proved to be sufficient to remove the impurities from the solution in the form of an easily filterable brown lump of solid. Cellulose added at this point was silylated to a completely soluble product. This ''sweetened'' procedure provides a more economical synthesis of silyl cellulose than the silylations with BSA.

Trimethylsilyl Ethyl Cellulose

Klebe and coworkers successfully made modification of cellulose derivatives by replacing residual hydroxy protons by silyl groups. It has been possible to trimethylsilylate under mild conditions an ethyl cellulose containing an average of 0.7 - 0.8

hydroxyl groups per anhydroglucose unit. The product which contained 0.6 - 0.7 trimethylsilyl group per ring was insoluble in alcohols but soluble in aliphatic hydrocarbons in contrast to the unsilylated ethyl cellulose.

Substitution With Other Silyl Groups

The two methods developed by Klebe and coworkers for trimethylsilylating cellulose provided general routes to other silyl celluloses as well. Reaction of cellulose with N-(dimethyl-γ-cyanopropylsilyl)acetamide in N-methylpyrrolidone or with the corresponding silyl chloride in pyridine allowed the substitution of 80-90% of the hydroxyl protons by dimethyl-γ-cyanopropylsilyl groups. In the same fashion, an average of 2.5 of the three hydroxyl protons of the anhydroglucose unit was replaced by dimethylphenylsilyl and methyldiphenylsilyl groups, whereas substitution with triphenylsilyl groups did not prove to be possible by either method.

Silylation With Hexamethyldisilazane

Harmon and coworkers (20,21) have developed a method for persilylation of cellulose using hexamethyldisilazane as the silylating reagent and formamide as solvent. Cellulose dissolves completely in formamide by heating at 70°C for 1-2 hours giving a homogeneous clear viscous solution. Excess of hexamethyldisilazane was added to the solution and the reaction mixture was heated at 70-80°C for 2 hours when the silylation was completed. The product was purified by repeated washings with anhydrous acetone. The silyl cellulose, obtained, contained 22.1% silicon indicating persilylation (3 silyl groups per anhydro glucose unit).

Nagy and coworkers (22) have studied the effects of a number of solvents in the silylation of cellulose with hexamethyldisilazane. The silylation of cellulose with hexamethyldisilazane was possible in the presence of some solvents, e.g., pyridine, dimethyl sulfoxide and dimethyl formamide, which rapidly formed unstable complexes which initiated the silylation reaction. Other solvents, e.g., acetonitrile, nitromethane or nitrobenzene, did not form complexes or formed them slowly, and had no effect on the silylation reaction.

Silylation With Trimethylsilylacetamide

Bredereck and coworkers (23) have reported persilylation of cellulose using trimethylsilylacetamide. Trimethylsilyl cellulose containing 22.1% silicon was obtained by melt polymerization at 170-180°C for 6 hours.

Properties of Silyl Celluloses

The properties of the silyl cellulose derivatives have been studied extensively by Klebe and coworkers (18). The proportion of the silyl groups in the silyl polymer ranges from about 55 wt-% for trimethylsilyl cellulose to about 75 wt-% for methyldiphenylsilyl cellulose. It is not surprising then, that the properties exhibited by the various silyl celluloses are largely determined by the nature of the silyl substituents.

Trimethylsilyl Cellulose. The polymer is colorless and is soluble in aromatic and chlorinated solvents and also in a number of aliphatic hydrocarbons. It is insoluble in alcohols, ketones, esters, nitriles, and other polar solvents. Solution-cast films are clear, flexible, and moderately strong (Table III). The polymer does not melt. Decomposition starts around 300°C in air. Electrical measurements show a low dielectric loss (Table IV). The high corona resistance is noteworthy. It is characteristic of polymers with a relatively high silicon content.

Table III. Silyl Celluloses

Substituent	Substitution (%)	Tensile strength psi	Elongation to break %
Me_3Si	89	4500	30
$Me(C_6H_5)_2Si$	83	-	-
$NC(CH_2)_3Me_2Si$	88	1500	198
$\frac{NC(CH_2)_3Me_2Si}{Me(C_6H_5)_2Si} = \frac{1}{3}$	~85	3200	20
$\frac{NC(CH_2)_3Me_2Si}{Me(C_6H_5)_2Si} = \frac{1}{1}$	~85	2500	30
$\frac{NC(CH_2)_3Me_2Si}{Me(C_6H_5)_2Si} = \frac{3}{1}$	~85	2200	100

Trimethylsilyl cellulose has relatively high hydrolytic stability. Polymer samples left in ambient air were completely soluble in benzene up to three months. Even after 3 hours in boiling water, the polymer was nearly completely soluble in benzene. After several days in boiling water, 98% of the silyl groups were gone. However, the shape and clarity of the film sample was essentially unchanged after this treatment.

Table IV. Electrical Data on Silyl Celluloses

Substituent	Substitution (%)	Frequency cps	Tan δ	Dielectric Constant
Me_3Si	89	60	0.0007	2.8
		10^3	0.0016	2.8
		10^4	0.002	2.8
		10^5	0.003	2.8
$NC(CH_2)_3Me_2Si$	88	60	0.169	14.5
		10^3	0.0192	14.1
		10^4	0.0315	13.8
		10^5	0.0811	12.7

The hydrolytic stability of trimethylsilyl cellulose is particularly remarkable in view of the fact that polytrimethylsilyl vinyl ether is solvolyzed by methanol to polyvinyl alcohol under mild conditions (2). Whereas repeated precipitations and prolonged treatment of trimethylsilyl cellulose with methanol had no effect on this polymer. The bonding of the silyl groups is quite similar in these two polysilylalkyl ethers. The enhanced hydrolytic stability of silyl cellulose must be attributed to steric hindrance.

In order to determine the relative hydrolytic stability of a number of different silyl celluloses, weighed film samples of 3 mils thickness were subjected to water or 1 atm. of water vapor (above 100°C) at different temperatures for certain lengths of time and then reweighed after thorough drying at 200°C/0.2 mm for 15 hours. The extent of hydrolysis that had occurred was calculated from the weight loss of the film. Figure 2a shows that 50% of the trimethylsilyl groups were lost by hydrolytic cleavage after 10 hours in water or water vapor of 100°C. Exposure to water of 70°C or water vapor of 120°C leads to 50% hydrolysis after 80 hours. After 80 hours exposure to water vapor of 150°C or water of 27°C, only 7% and 2% hydrolysis, respectively, results.

The rather slow rate of hydrolysis in water vapor at high temperature, which was also observed with cellulose substituted with dimethyl-γ -cyanopropylsilyl, dimethylphenylsilyl, and methyldiphenylsilyl groups is peculiar, even considering the lower water content of the vapor, which is at 150°C about one-fifth of that at 100°C.

Dimethylphenylsilyl Cellulose. This polymer resembles trimethylsilyl cellulose in all properties investigated. Monophenyl substitution on silicon does not lead to enhanced hydrolytic stability, as shown in Figure 2b.

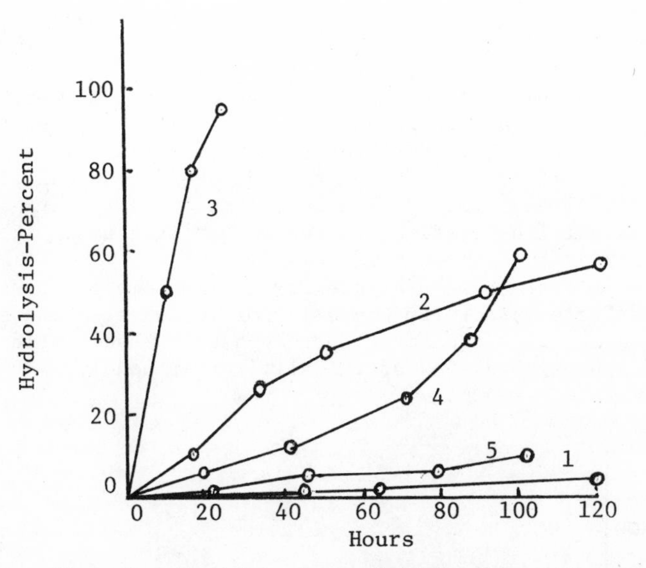

Figure 2a. Hydrolysis of trimethylsilyl cellulose: (1) in water, 27°C; (2) in water, 70°C; (3) in water, 100°C; (4) 1 atm. of water vapor, 120°C; (5) 1 atm. of water vapor, 150°C

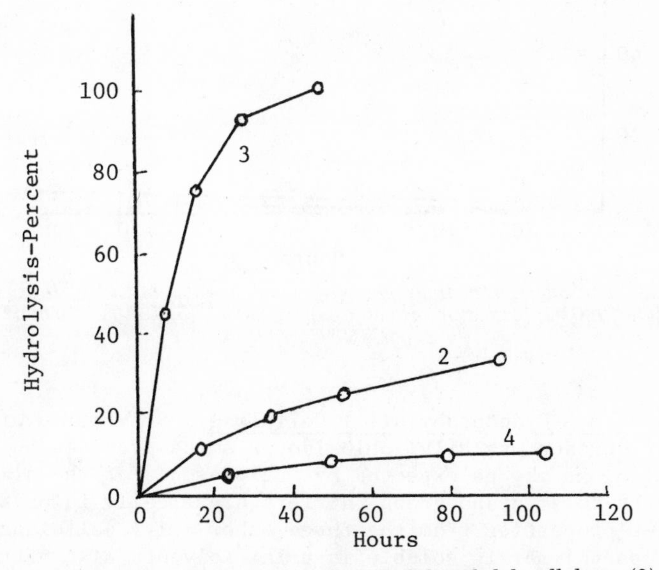

Figure 2b. Hydrolysis of dimethylphenylsilyl cellulose: (2) in water, 70°C; (3) in water, 100°C; (4) 1 atm. of water vapor, 120°C

Methyldiphenylsilyl Cellulose. Like the trimethylsilyl and the dimethylphenylsilyl analog, methyldiphenylsilyl cellulose is a colorless polymer, soluble in aromatic solvents, in chlorinated hydrocarbons and in pyridine. The polymer does not melt. Decomposition takes place at around 350°C. Cast films are rather stiff, like those of the other two members of the silyl cellulose family, but less flexible. Sharp creasing does not break films of trimethylsilyl and dimethylphenylsilyl cellulose, whereas the methyldiphenylsilyl derivative fractures when treated in this way. The outstanding property of methyldiphenylsilyl cellulose is its hydrolytic stability (Figure 2c). Film samples appeared entirely unaffected after 120 hours in boiling water. The slight weight loss after this treatment corresponded to hydrolytic cleavage of approximately 5% of the silyl ether bonds.

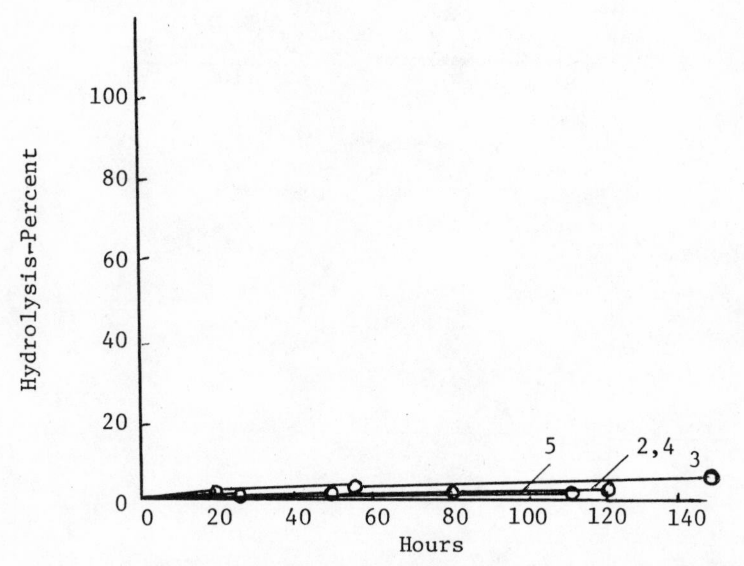

Figure 2c. Hydrolysis of methyldiphenylsilyl cellulose: (2) in water, 70°C; (3) in water, 100°C; (4) 1 atm. of water vapor, 120°C; (5) 1 atm. of water vapor, 150°C

Dimethyl-γ-Cyanopropylsilyl Cellulose. Silylation with dimethyl-γ-cyanopropylsilyl chloride or amide provides this polymer which, as may be expected by the presence of the alkyl chain and the polar cyano group, differs drastically in physical and solution properties from the three other silyl celluloses. The colorless polymer is soluble in polar solvents like nitriles and ketones. It is insoluble in aliphatic and aromatic hydrocarbons. Cast films are slightly opaque and very soft. They can be shaped to some degree by hand without tearing. The hydrolytic stability of this polymer is somewhat lower than that of trimethylsilyl cellulose as shown in Figure 2d, although the hydrolysis under ambient conditions is insignificant.

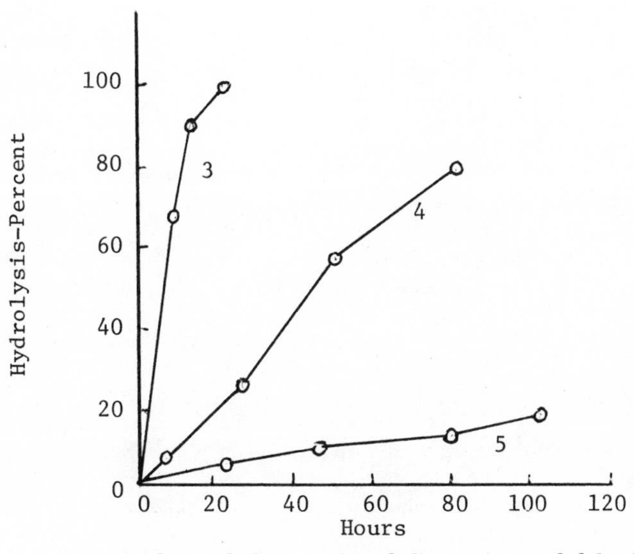

Figure 2d. Hydrolysis of dimethyl-γ-cyanopropylsilyl cellulose: (3) in water, 100°C; (4) 1 atm. of water vapor, 120°C; (5) 1 atm. of water vapor, 150°C

Cosilylated Cellulose. Cellulose has been cosilylated with dimethyl-γ-cyanopropylchlorosilane and methyldiphenylchlorosilane in various relative proportions. Cast films of the resulting polymers rich in methyldiphenylsilyl groups are nearly clear whereas a large proportion of the polar silyl group results in opacity. Tensile and elongation data (Table III) of these mixed substituted silyl celluloses also reflect the apparently additive effect of the two different substituents on the properties of the polymers.

Literature Cited

1. Klebe, J.F., J. Polym. Sci. B, (1964), 2, 1079.
2. Murahashi, S; Nozakura, S.; and Sumi, M; J. Polym. Sci. B, (1965), 3, 245.
3. Ebelman, J.J., Compt. Rend., (1844), 18, 1202; 19, 398.
4. Kalinin, M.N., Compt. Rend. Acad. Sci. URSS, (1940), 26, 365.
5. Martin, G., J. Chem. Soc., (1914), 105, 2860.
6. Patnode, W.I., U.S. 2,306,222 (Dec. 22, 1942).

7. British Thomson-Houston Co., Brit. 575,696 (Feb. 28, 1946).
8. Hyde, J.F., U.S. 2,413,050 (Dec. 24, 1946).
9. Jullander, I., ''Studies of Nitrocellulose'', pp. 109-117, H.K. Lewis and Co. Ltd., London, 1945.
10. Schuyten, H.A.; Weaver, J.W.; Reid, J.D.; and Jurgens, J.F.; J. Amer. Chem. Soc., (1948), 70, 1919.
11. Schuyten, H.A.; Weaver, J.W.; Reid, J.D.; U.S. 2,562,955 (Aug. 7, 1951).
12. Hunter, M.J., U.S. 2,532,622 (Dec. 5, 1950).
13. Ivanov, N.V.; Rogovin, Z.A.; and Chih, N.W.; Izv. Vysshikh Uchebn. Zavendenii, Khim. i Khim. Teknol., (1965), 8(1), 124-6.
14. Predvoditelev, D.A.; and Rogovin, Z.A.; Vysokomol. Soedin., Ser. B., (1967), 9(8), 611-12.
15. Klebe, J.F.; U.S. 3,418,313 (Dec. 24, 1968).
16. Klebe, J.F.; U.S. 3,418,312 (Dec. 24, 1968).
17. Finkbeiner, H.L.; and Klebe, J.F.; U.S. 3,432,488 (March 1, 1969).
18. Klebe, J.F.; and Finkbeiner, H.L.; J. Polym. Sci., Part A-1, (1969), 7(7), 1947-58.
19. Klebe, J.F.; Finkbeiner, H.L.; and White, D.M.; J. Amer. Chem. Soc., (1966), 88(14), 3390.
20. Harmon, R.E.; De, K.K.; and Gupta, S.K.; Carbohyd. Res., (1973), 31, 407.
21. Harmon, R.E.; De, K.K.; and Gupta, S.K.; Starke, (1973), 25(12), 429.
22. Nagy, J.; Kuszmann, A.B.; Palossy, K.B.; and Hegedus, E.Z.; Makromol. Chem., (1973), 165, 335-8.
23. Bredereck, K.; Strunk, K.; and Menrad, H.; Makromol. Chem., (1969), 126, 139-46.

5

Utilization of the Novel Reaction of Cellulose with Amic Acids to Produce Cellulose Derivatives Containing Carboxylic Acid Groups

THOMAS C. ALLEN
New Products and Processes Research, American Enka Co., Enka, N.C. 28728
JOHN A. CUCULO
Department of Textile Chemistry, North Carolina State University, Raleigh, N.C. 27607

Introduction

One of the main reasons for the widespread use of cellulose has been the abundance of hydroxyl groups and their ability to react with a variety of compounds to form cellulose derivatives. These derivatives exhibit selected desired properties, depending on the new functional group added. Although the modification of cellulose has been studied and practiced commercially for several years, new reactions and new derivatives with new properties are still being investigated and much of the chemistry involved is not completely understood. One of the most studied functional groups is carboxylic acid. In a recent review the methods of preparation and the properties of these carboxylic acid containing derivatives were discussed (1). Some of these properties, which are under active investigation today are:

- salt formation
- solubility of alkali metal and amine salts in water
  - fiber formation
  - films, coatings, and binders
- ion-exchange
- cationic dye attraction
- water absorbency
- soil release
- opening of structure for subsequent reactions
- new site for chemical reactions
  - crosslinking
  - monomeric
  - grafting

One class of such derivatives, the cellulose semi-esters, has achieved little commercial significance, even though it was first investigated over thirty years ago.

This paper is concerned with the novel reaction of cellulose with amic acids prepared by the reaction of polycarboxylic acid anhydrides with ammonia either beforehand or even in situ in the treating solution. This reaction, first reported by Cuculo (2,3,4) and later studied for generality (5,6) offers the unique features of using relatively inexpensive and readily available materials in a simple aqueous pad-bake technique with short reaction times. Furthermore, by choosing different amic acids such as succinamic, maleamic, and phthalamic, aliphatic, unsaturated, and aromatic derivatives can be obtained.

Three major products can be expected from the reaction of cellulose with amic acids: (a) the cellulose half-acid ester, (b) the crosslinked cellulose diester, and (c) the cellulose half-amide ester.

$$\text{Cellulose-OH} + \text{NH}_2\text{C(O)RCO}_2\text{H} \longrightarrow$$

(a) $\text{Cellulose-OC(O)RCO}_2\text{H}$

(b) $\text{Cellulose-OC(O)RC(O)O-Cellulose}$

(c) $\text{Cellulose-OC(O)RC(O)NH}_2$

The cellulose half-acid ester was shown to be present in relative large amounts as evidenced by tests for the carboxylic acid. Comparison of the free carboxyl value with the total carboxyl value showed that one or both of the other products was present. The half-amide ester was believed to be a minor product due to the copious evolution of ammonia during the reaction and very low nitrogen analysis of the treated fabric (2,5).

The object of this work is to explore the amic acid reaction to determine if the many variables in the process can be controlled to maximize the formation of the cellulose half-acid ester and to increase the extent of reaction so that the properties peculiar to the carboxylic acid group can be realized.

Experimental

Reactants. Succinamic acid was the principal reactant in this study due to its demonstrated reactivity and its relatively high solubility (2,5). It was prepared by dissolving as much as possible of 500 grams of commercial grade succinic anhydride in one liter of acetone. The solution was then cooled to 10-20°C, leaving the undissolved anhydride as a precipitate in acetone. Then 300 ml. of a 29% aqueous ammonia solution was added as fast as possible but slow enough so that no boiling occured. The solution was cooled to 0-10°C with constant stirring until the amic acid had precipitated. After filtering under vacuum, the crystals were washed with acetone in the funnel with the vacuum off for five minutes before sucking dry. This wash procedure was repeated two times. The crystals were spread on a tray and allowed to dry. A 75% yield was obtained. Purity was determined by melting point (154-156°C), neutralization equivalent, and nitrogen analysis. Maleamic acid was made by the same procedure. Other amic acids were purchased and used as received.

The woven fabric used in most of the experiments in this investigation was a 100 per cent viscose rayon satin weave, 120 x 40, 150 denier x 75 denier with a yarn weight of 2.75 ounces per square yard. It was desized, scoured, and dried before treatment. A cotton fabric of 78 x 78, 30/1 x 40/1 3.5 oz./yd.^{2}, satin weave construction was used in some experiments. The fabric was tentered to 80 x 80 standard print construction after desizing, scouring, and bleaching.

Fibers and wood pulp were formed into nonwoven mats by dispersing in water and filtering on a fiberglass screen in a Buchner funnel under vacuum. About 2.5 grams of dry fibers on a 250 mm. funnel gave a mat thin and uniform enough for the treatment desired. Fibers treated by this process were:

Viscose Rayon- Beaunit Fibers, Code 212, bright regular rayon staple, 1.5 denier, cut 1 9/16 inch.

Viscose Rayon - Never Dried- ITT Rayonier Co., regular rayon staple fiber never dried, 1.5 denier, 3/8 inch staple length, finished with Arquad 2C75 as anti-mildew agent.

Cellulose Acetate- Celanese Corporation, Lot 44R, Bale 26294, bright staple, 3.0 denier, 1 9/16 inch cut.

Wood Pulp- Rayacord XF-ITT Rayonier Co., southern pine, sulfite bleached, approximately 97% alpha cellulose, cuene intrinsic viscosity of 9.2.

Never Dried Cotton-Cotton bolls stored in water were obtained from Cotton, Inc. and cut open under water to remove fibers from seeds with a razor blade.

All water used in the treating solution as well as in all purification and analytical procedures was distilled and then deionized by passing through a Barnstead Type HN, Multiple Bed demineralizer. All catalysts and solvents were purchased and used as received.

Treating Solution. At a 2 to 1 pick-up ratio, an 18% (0.0154 molar) solution of succinamic acid in water gives a 0.5 to 1 molar ratio of succinamic acid to anhydroglucose unit. This 18% solution and higher concentrations must be kept warm, at least 40°C, to keep the succinamic acid in solution. The only concentrations used lower than 18% were 9% or 0.25/1 molar ratio SAA/AHGU at 2/1 PUR. This concentration did not require heat for solution or precipitation but heat was useful to reduce the time needed for complete solution. The general procedure followed in all bath make-up was to weigh the succinamic acid into a beaker, add warm water to about 90 per cent total weight, stir until all the succinamic acid was dissolved (on a stirring hot plate), add catalyst if any, and bring to total weight with water from a squeeze bottle.

Fabric Treatment. Most of the fabric treatments were done by dipping the weighed fabric into the solution, which was kept warm to prevent precipitation, for one minute and then removing the fabric and squeezing it by hand until the weight indicating the desired pick-up ratio was obtained. This dip procedure appears to be quite satisfactory due to the reproducibility of various experiments and the uniformity of treatment indicated by dyeing the treated fabric with indicator dye (Geigy Fiber Indicator Syn and Identification Stain No. 1 New), and by dipping the fabric in water containing a small amount of methylene blue, squeezing to a desired pick-up ratio, and drying.

Nonwoven fabrics had to be placed between a fiberglass screen before dipping to prevent dispersion of the fibers. After dipping, most of the excess treating solution was squeezed out. The fabric was then removed from the screen and squeezed to the desired pick-up ratio.

Fabric Heat Treatment. The treated fabric was either air dried by hanging on a line at room temperature, heat dried by placing in an oven and observing the temperature with a thermocouple attached to the fabric, or cured immediately. Curing or baking was accomplished by suspending the sample vertically in a large forced air oven of internal size 36 x 25 x 19 inches.

Purification. As soon as possible after drying, the fabric was placed in water to prevent further reaction and to remove unreacted material. Successive 15 to 30 minute washes at room temperature of water, acetone, water until colorless, 1N H_2SO_4 to convert all acid salts to the free acid, and 1N H_2SO_4 were given the fabric. The cured fabric was usually colored from light to dark brown, depending on the time and temperature of cure. Some, but not all, of the color was removed by the water, acetone, and sulfuric acid in the above procedure. Several fast (5 to 15 minutes) washings with water then followed. The fabric was then cut with scissors into small pieces and put into an automatic blender containing water. The blender was turned on high speed for 15 to 20 seconds or until the bits of fabric were reduced to fibers. The fibers were then filtered under vacuum on a Buchner funnel. The filtrate was discarded and the fibers were washed with water until a color change from yellow to blue occurred when two drops of a 0.1% solution of bromcresol purple indicator in 95% ethanol and then one drop of a 0.1 N sodium hydroxide solution were added to 100 ml. of the filtrate. After all the unattached acid was out of the fibers, as evidenced by the indicator color change above, the water remaining in the fibers was exchanged with acetone. The fibers were then dried at 60 ± 5°C for one to two hours under vacuum in a heated vacuum dessicator.

Pendant Carboxyl. A modification of the calcium acetate method described by Cuculo (2) was used to determine the pendant carboxylic acid groups attached to the cellulose. A blank was determined by titrating 50 ml of the original 0.133N calcium acetate solution in the same manner. A further blank value was determined for untreated viscose rayon fabric, which was washed and processed exactly the same as the treated fabric. Both blanks were subtracted from all treated sample values in the tables. The results expressed as milliquivalents of half-acid ester per gram of sample were calculated as follows:

meq. NaOH used - meq. HCl used - meq. calcium acetate blank -meq. untreated substrate blank = meq. NaOH consumed for pendant COOH = meq. COOH

$$\frac{\text{meq. COOH}}{\text{N NaOH}} = \text{ml. NaOH consumed for pendant COOH}$$

$$\frac{2 \times \text{N NaOH (ml. NaOH consumed + ml. correction factor)}}{\text{wt. of sample}} =$$

actual meq. COOH per gram of sample

It was found about half way through this work that the test method described was satisfactory at low amounts of pendant carboxyl content. However, as the amount of cellulose half-acid ester increased in relation to the constant calcium acetate concentration, the exchange of calcium ion decreases, resulting in a low determination for pendant carboxyl. Finally, a method suggested by Cramer involving the titration of several sample sizes of varying carboxylic acid content (7)was applied. A plot of sample size vs carboxyl content and extrapolation to zero sample size gives the true carboxylic acid content. A further plot of the true carboxyl content minus the found carboxyl content for a standard sample size (one gram in this work) versus the found carboxyl content for a variety of different carboxylic acid containing samples gives the correction factor in milliters of NaOH used for any subsequent sample.

Total Ester. The total carboxyl test method depends on saponification which converts the carboxylic ester groups attached directly to the cellulose molecule to carboxylic acid groups, with back titration determining the amount of total carboxylic acid groups present, both those that came from the ester linkage and those present as pendant carboxyl. The method chosen was the Eberstadt method as modified by Genung and Mallatt(8). A further modification used here was that 50 ml. each of 75% ethanol and 0.5 N NaOH was used instead of the 40 ml. used by Genung and Mallatt. In later experiments, the 75% ethanol, which was deemed questionable in value by Genung and Mallatt (8) was replaced with water with no change of results. The water is more likely to swell the esters studied here than ethanol. All flasks containing the sample stood in a tray of water for 48 hours. The tray of water was used when fluctuations of results appeared on repeat analyses of the same sample and was attributed to temperature fluctuations. Blanks containing alcohol or water and NaOH were run with each set of samples. Calculations for percent total COOH are the same as for percent free COOH except the factor of two is not needed here. Results were expressed as milliequivalents of total ester per gram of sample by subtracting the pendant from the total carboxyl value.

Calculations. The results were expressed as milliequivalents per anhydroglucose unit (meq./AHGU), calculated by the general method of Allen (9). The specific procedure used was:

H = milliequivalents of half-acid ester per gram of sample

T = milliequivalents of total ester per gram of sample

A = milliequivalents of half-amide ester per gram of sample

C = milliquivalents of cellulose per gram of sample

and for the reaction of succinamic acid

$$0.100\,H + 0.084T + 0.099A + 0.162C = 1.00$$

H, T and A were obtained as previously described so C can be calculated.

then,

milliequivalents of half-acid ester per anhydroglucose unit or pendant D.S. = $\frac{H}{C}$

and, milliequivalents of total ester per anhydroglucose unit or total D.S. = $\frac{T}{C}$

and, milliequivalents of half-amide ester per anhydroglucose unit or half-amide ester D.S. = $\frac{A}{C}$

Half-Amide Ester Determination. Nitrogen content was determined by the Kjeldahl method (9). The results were expressed as milliequivalents of amide per gram of sample, as follows:

$$\text{meq. amide per gram of sample} = \frac{\text{Meq. HCl used for sample - meq. HCl for blank}}{\text{weight of sample}}$$

Results and Discussion

Reaction Method. A first step in this investigation was to determine if some other method of reaction other than the pad-bake method used by Cuculo (2) offered more promise in obtaining the desired objective of increasing the D.S. and maximizing the formation of the cellulose half-acid ester. One method investigated was the heating of a solution of succinamic acid with the cellulose for a period of time. This is the method usually used in cellulose esterification where the anhydride or the acid is the reactant. The high ratio of succinamic acid to cellulose was used in anticipation that the Mass Action Law would favor the formation of the monoester. No catalyst was present in this or the other two methods.

The other method chosen was the placement of the cellulose in molten succinamic acid held just above the melting point at approximately 160°C for ten minutes.

The results of the three reaction methods are shown in Table I.

Table I: Effect of Reaction Method in the Succinamic Acid-Rayon Fabric Reaction

Reaction Method	Moles Succinamic Acid/AHGU	Total Ester Meq./AHGU	Half-Acid Ester Meq/AHGU	Half-Amide Ester Meq/AHGU	Cross links Me/AGU
Pad-Bake	0.872	0.316	0.272	0.013	0.015
Melt	34.6	0.283	0.144	0.048	0.045
Reflux	16.2	0.195	0.077	0.028	0.045

Reaction Conditions:

Pad-Bake: 18% aqueous succinamic acid solution, 3.5 pick-up ratio (PUR), cure 150°C 10 minutes.

Melt: dip fabric in molten succinamic acid (approx. 160°C) for 10 minutes.

Reflux: 58.5% succinamic acid in N,N-Dimethylformamide (DMF), reflux 150-160°C for 75 minutes.

The first surprising aspect of the amic acid reaction is immediately seen from the total ester results. Even though the pad-bake procedure has less succinamic acid available to the cellulose, more ester groups are formed than in the other two methods. Furthermore, the desired objective of miximizing the formation of the cellulose half-acid ester is significantly greater in the pad-bake procedure where 84% of the ester groups contain free carboxyl on the other end compared to 51% and 39% for the melt and reflux methods, respectively. A ready explanation at this point is that the water used in the pad-bake procedure allows a more complete penetration and opening of the fiber structure. Further examination of the results show that the pad-bake procedure gives only about 4% of the total ester groups containing the amide group on the other end, while the melt and reflux procedures contain respectively about 17% and 14% amide groups in relation to ester links. Finally, only about 7% of the carboxylic acid group formed reacted further to form ester crosslinks in the pad-bake method while in the melt and reflux methods, the extent of further reaction was 24% and 37% respectively.

Somewhat surprising low extent of reaction and high percentage of crosslinking in the reflux procedure led to a further exploration of this reaction method for succinamic acid in regard to solvent type. In Table II a number of solvents that are

Table II: Effect of Solvent Type in Reflux Reaction of Succinamic Acid with Rayon Fabric

Solvent	Total Ester Meq/AHGU	Half-Acid Ester Meq/AHGU	Half-Amide Ester Meq/AHGU	Crosslinks Meq/AHGU
DMF	0.195	0.077	0.028	0.045
Formamide	0.109	0.041		0.034
Dimethyl Sulfoxide	0.116	0.070		0.023
Water	0.067	0.051		0.008
Pyridine	0.230	0.181	0.003	0.023
Succinic Anhydrife in Pyridine	0.643	0.380		0.132

Reaction Conditions:

0.5 molar succinamic acid in each solvent, heat with fabric 150-160°C (or reflux if boiling point of solvent is less than 150°C) for 75 minutes except 2 1/2 hours with H_2O.

generally regarded to have some swelling effect on cellulose are compared. Some difference in reaction temperature was present due to boiling point differences. Formamide and dimethly sulfoxide gave somewhat less reaction than DMF but DMSO did give a higher percentage of carboxylic acid groups (60% compared to 39%)

Another quite surprising result occurs with water as the solvent. Although the extent of reaction is only about 1/3 that of DMF and 1/2 that of formamide and DMSO, such significant esterification is not generally expected in the presence of water. Also, the reaction temperature was only 100°C. The percentage of free acid groups, compared to total ester groups, is about the same as in the pad-bake procedure.

Pyridine is regarded as a classical esterification solvent, being extensively used in studies of the reaction of various acids, acid chlorides, and acid anhydrides with cellulose. Although the extent of reaction is significantly greater than with the other solvents studied, it is still somewhat lower than in the aqueous pad-bake method from Table I, and about 1/3 as much as that with succinic anhydride in pyridine. The latter comparison indicates a significantly lower rate of reaction of succinamic acid compared to succinic anhydride. However, with the amic acid, only 12% of the free acid groups reacted further to give crosslinks compared to 26% with the anhydride. On the other hand, 76% of the total ester groups are in the form of the half-acid ester in the amic acid reaction, compared to 59% in the anhydride reaction. The lower extent of crosslinking with anhydride in pyridine solvent is generally explained by the formation of the pyridinium salt of the free acid group. However, the use of succinamic acid further reduces crosslinking, possibly due to the formation of the ammonium salt and again possibly due to a unique mechanism of reaction.

Since solvent seems to have some effect on the extent of half-acid ester formation in the reflux reaction, the pad-bake procedure was examined for the effect of applying the amic acid from materials other than water. Table III shows a number of surprising results. First, the effect of pyridine in giving a high amount of free acid in the reflux reaction is essentially nonexistent in the pad-bake reaction. Second, the relation of formamide and DMF is reversed. Formamide gives an extent of reaction only somewhat less than that of water while DMF gives very little reaction in the pad-bake process compared to the 0.2 D.S. for the reflux method. An effect of the amide proton is suggested by the fact that N-methyl formamide gives about the same result as formamide. The relationship is seen with the pair of N-methyl pyrrolidone and 2-pyrrolidone, except 2-pyrrolidone does give less reaction than N-methyl formamide. The extent of reaction with DMSO compared to formamide is also lower than expected from the reflux reaction. Yet even less reaction is obtained on going to the sulfone group and a cylic structure in the form of tetramethylene sulfone.

In Table IV, variations on the use of formamide and dimethylformamide are explored. Increasing the amount of succinamic acid in DMF by four times increases the extent of reaction more than four times but still not to the extent of that obtained in water. Addition of a catalyst to the lower concentration of succinamic acid in DMF also increases the extent of reaction. Furthermore, the combination of water and DMF as solvent increases the extent of reaction almost to that obtained with

Table III: Effect of Solvent in the Pad-Bake Reaction of Succinamic Acid with Rayon Fabric

Solvent	Total Ester Meq/AHGU	Half-Acid Ester Meq/AHGU	Half-Amide Ester Meq/AHGU	Crosslinks Meq/AHGU
Water	0.161	0.113		0.024
DMF	0.020	0.024		0.000
Formamide	0.179	0.120		0.030
N-Methyl Formamide	0.134	0.114		0.010
2-Pyrrolidinone		0.050		
N-Methyl Pyrrolidinone		0.022		
Pyridine*		0.038		
Dimethyl Sulfoxide	0.101	0.075	0.001	0.013
Tetramethylene Sulfone		0.019		

Reaction Conditions:
9% succinamic acid, 2.0 PUR, cure 150°C 8 minutes.

* Small amount of water added for solubility.

water alone. However, the mixture of water and formamide offers no improvement over formamide alone. The obvious conclusion from these results is that water serves some unique function in the pad-bake reaction of succinamic acid with cellulose, duplicated by only formamide and possibly sulfamic acid. This function is not as pronounced in reflux reactions.

One could argue that the optimum solvent, time, temperature or catalyst has not been used in the few reflux experiments performed here and that more experimental work could produce better results. The same arguments could also be applied to the melt method, particularly with regard to the accessibility of the reactants to each other. In addition, a pre-swollen cellulose may be more reactive. Thus, a woven viscose rayon fabric wet with an 18% solution of succinamic acid (to a 4.0 pick-up ratio) was placed in the molten succinamic acid at 160 to 170°C for 10 minutes. Surprisingly, this procedure gave less reaction than

Table IV. Continued Investigation of Solvent Effect in the Pad-Bake Reaction of Succinamic Acid with Rayon Fabric

Solvent	Total Ester Meq/AHGU	Half-Acid Ester Meq/AHGU	Half-Amide Ester Meq/AHGU	Crosslinks Meq/AHGU
DMF/Water 50/50	0.155	0.125		0.015
Formamide/Water 50/50	0.137	0.101	0.002	0.018
DMF + Sulfamic Acid Catalyst (6% on wt. of succinamic acid)	0.152	0.101	0.002	0.026
DMF but 36% succinamic acid	0.192	0.155	0.002	0.018
Water but 36% succinamic acid	0.326	0.282		0.022

Reaction Conditions: 9% succinamic acid except where noted, 2.0 PUR, cure 150°C 8 minutes.

that of just putting the dry fabric in the melt. The water in the fabric should have aided in the diffusion of the succinamic acid into the fabric and it should not have interferred with the reaction since it was evaporated away very rapidly. Decomposition of amic acids by water is very common (10)and could have occurred in this case. Other experiments, such as the use of a non-aqueous swelling solvent or the use of solvent-exchanged fabric, may increase the D.S. However, it was concluded that an investigation of the variables in either of these two processes offered no more promise of obtaining the desired objectives than the study of the many variables in the pad-bake process, which as already seen was much more promising in the initial screens.

The next step was to then investigate the variables in the aqueous pad-bake process to determine those that had the greatest effect on both the extent of esterification and the subsequent crosslinking.

The unusual effect of the presence of water was seen again when the drying conditions were investigated. It was reasoned that air drying after impregnation of the fabric could give a better distribution of the reactant, leading possible to more extensive and uniform reaction. As seen in Table V, both the total ester and carboxyl content decrease significantly after even one hour of air drying before the heat treatment. The D.S. decreases even more after 3 hours at ambient temperature before

Table V. Effect of Drying Conditions on Succinamic Acid-Rayon Fabric Reaction

Drying Conditions	Total Ester Meq./AHGU	Half-Acid Ester Meq./AHGU	Crosslinks Meq./AHGU
None- Cure Wet	0.154	0.133	0.011
Air Dry 1 Hour	0.121	0.102	0.010
Air Dry 3 Hours	0.065	0.067	0.000

Reaction Conditions: 9% aqueous succinamic acid solution 2.0 PUR, Cure 150°C 12 minutes.

curing. The decrease in D.S. continues and levels off at around 24 hours of air drying. In fact, after 24 hours at ambient temperature between treatment and curing, it takes 15 minutes at 150°C to obtain the same extent of reaction with a 9% by weight succinamic acid solution as obtained in 6 minutes at 150°C with no time between treating and curing. As the amount of succinamic acid in the bath increases, the time differential increases further. Yet significant reaction is obtained at low temperatures in long periods of time such as: 125°C total D.S. in 1 hour = 0.19; 125°C total D.S. in 16 hours = 0.36; 100°C total D.S. in 24 hours = 0.18; ambient temperature total D.S. in 3 months = 0.05. The extent of reaction differential is not observed if the fabric is dried at an elevated temperature (120°C or greater) and then placed at ambient temperature for extended periods of time before curing. It is possible that the presence of water during the heat treatment allows some sort of stereospecific hydrogen bonded complex of the amic acid and cellulose to be set up in preference to the loss of water to form an imide. The role of water as a transport agent is also a possibility.

Hydrolyis of amic acids to the diacid is a distinct possibility, so the stability of the treating solution was investigated. Using a 18% succinamic acid solution no change was seen in extent of reaction on fabric treated in a fresh solution compared to a solution maintained at 60 to 70°C for two hours. Furthermore, fabric was maintained in the solution for one or fifty minutes, with or without sulfamic acid catalyst (Table VI). However, an interesting effect is seen in this table in comparison of the amount of total ester obtained with catalyst present compared to no catalyst. The use of catalyst actually reduces the total ester content while maintaining the half-acid ester content, thus reducing the amount of crosslinking.

With this lead in hand, a number of various compounds were investigated for possible catalyst activity, beginning with

Table VI: Effect of Time in Treating Solution in Reaction of Rayon Fabric with Succinamic Acid

Fabric Time in Bath (Min.)	Total Ester Meq./AHGU	Half-Acid Ester Meq./AHGU	Apparent Crosslinks Meq./AHGU
No Catalyst			
1	0.225	0.194	0.016
50	0.227	0.200	0.013
Catalyst			
1	0.195	0.208	0.000
50	0.202	0.209	0.000

Reaction Conditions: 18% aqueous succinamic acid, catalyst is 6% sulfamic acid on weight of succinamic acid, maintain treating solution at 50-60°C, 2.0 PUR, cure 150°C 6 minutes.

acids and extending to bases, salts, and amides. As seen in Table VII some compounds of each class gave a slight increase in the extent of reaction, but the only significant catalyst effect was the confirmation of the reduction of crosslinking by sulfamic acid. Ammonium sulfamate behaved similarly while sodium sulfamate was less effective. Interesting comparisons are the increase in reaction with maleic acid while fumaric, succinic, citric, and tartaric acids were the same as the control. Ammonium sulfate gave no increase in reaction or decrease in crosslinking. At the same concentration, urea, thiourea formamide, and biuret showed no activity, compared with the 10% increase in total ester by guanidine hydrochloride.

The unique effects of ammonia as an additive, undoubtedly resulting in the formation of ammonium succinamate, and ammonium sulfamate in Table VII led to the exploration of other salts of succinamic acid. This concept is interesting from the possibility of reducing crosslinking similar to that with tertiary organic bases as solvent and catalyst in the anhydride reflux reaction (11), or the possibility of increasing the extent of reaction in the same manner as that of using the sodium and triethylammonium salts of polycarboxylic acids (12,13,14). Although in Table VIII the extent of reaction was actually decreased by the metal and primary amine salts and the extent of crosslinking was not reduced by the tertiary amine salts, a very significant increase in both total ester and pendant carboxyl was found with the triethylammonium salt. Although no reason for

Table VII: Effect of Catalyst in Succinamic Acid-Rayon Fabric Reaction

Catalyst	Total Ester Meq./AHGU	Half-Acid Ester Meq./AHGU	Apparent Crosslinks Meq./AHGU
None	0.245	0.198	0.024
Maleic Acid	0.272	0.236	0.018
Phosphoric Acid	0.264	0.232	0.016
Sulfamic Acid	0.208	0.208	0.000
Ammonia*	0.259	0.220	0.020
Monoammonium Phosphate	0.245	0.216	0.014
Ammonium Sulfamate	0.235	0.231	0.002
Sodium Sulfamate	0.251	0.222	0.015
Sodium Hypophosphite	0.245	0.228	0.008
Urea**	0.264	0.226	0.019
Guanidine HCl	0.299	0.184	0.057

Reaction Conditions: 18% aqueous succinamic acid, 2.0 PUR, cure 150°C 6 minutes, add succinamic acid, 90% of water, heat to 90°C, add catalyst and then remaining water, catalyst concentration is 1.08% except where noted.

*Concentration is 23.3% on weight of solution.
**Concentration is 5% on weight of solution.

the increased reaction of the triethylammonium salts of polycarboxylic acids was given in the previous work (12). The results here indicate that at least some reaction is taking place through the carboxylic acid group of the amic acid.

Table IX shows a very important discovery in meeting the stated objective of increasing the extent of reaction - the balancing of cure time and temperature with ratio of amic acid and water to cellulose. As the cure time is increased with increasing succinamic acid concentration, the total ester D.S. increases to over 1.0, quite a high value for a pad-bake reaction. Also, the half-acid ester content is lower at a value of around 0.65, indicating increased crosslinking.

Table VIII: Effect of Amine and Metal Salts on the Reaction of Succinamic Acid with Cellulose

Cation	Total Ester Meq./AHGU	Half-Acid Ester Meq./AHGU	Apparent Crosslinks Meq./AHGU
None	0.624	0.488	0.608
Ethanolamine	0.065	0.059	0.003
Ethylenediamine	-	0.022	-
Pyridine	0.636	0.472	0.082
Sodium	0.101	0.110	0.000
Calcium	0.106	0.092	0.007
None*	0.336	0.302	0.017
Triethylamine*	0.489	0.363	0.063

Reaction Conditions: 36% succinamic acid plus one equivalent of cation plus 3% sulfamic acid on weight of succinamic acid, 4.0 PUR, cure 150°C 18 minutes.

*2.0 PUR, cure 150°C 18 minutes.

The amount of carboxylic acid groups formed that reacted further to form crosslinks increased with cure time from a 5% to a 10% figure at the 6 and 9 minute cures to a 14, 18, 22, and 29% figure for the 18, 22, 33, and 170°C 22 minute cures respectively for the 54% succinamic acid solution. Table II showed that the amount of amide formation in the pad-bake reaction is small but the amount at a higher total ester content was of interest. It was found that the ratio of amide to crosslink actually decreased with extent of reaction. For example, the 54% succinamic acid solution, 150°C 22 minute cure sample had an amide D. S. of only 0.03 out of the 0.131 crosslink figure or a ratio of about 4 to 1 compared to the approximately 1 to 1 ratio at lower amic acid concentration and cure time in Table II.

In Table X, it is seen that the level off and actual decrease of D.S. with increasing succinamic acid concentrations does take place when wood pulp is used as the substrate, even at longer cure times. This result could be due to the less open structure of wood pulp, a concept that will be explored in greater detail later.

Another way to increase the ratio of amic acid to cellulose and the cure time is the use of repetitive treatments. Table XI

Table IX: Effect of Curing Conditions at Higher Ratios of Succinamic Acid to Cellulose in Amic Acid - Rayon Fabric Reaction

Moles Succinamic Acid/AHGU	Cure Time (min.)	Amic Acid Conc.%	PUR	Total Ester Meq/AHGU	Half-Acid Ester Meq/AHGU	Apparent Crosslinks Meq/AHGU
0.997	6	18	4.0	0.267	0.227	0.020
0.997	12	18	4.0	0.303	0.274	0.014
0.997	6	36	2.0	0.209	0.201	0.004
0.997	9	36	2.0	0.336	0.302	0.017
1.231	6	36	4.0	0.183	0.157	0.013
1.231	9	36	4.0	0.469	0.395	0.037
1.231	18	36	4.0	0.624	0.488	0.068
2.991	18	54	4.0	0.719	0.500	0.082
2.991	22	54	4.0	0.866	0.594	0.131
2.991	33	54	4.0	0.996	0.638	0.179
2.991	22*	54	4.0	1.177	0.653	0.262

Reaction Conditions: Viscose rayon fabric, cure 150°C
*170°C

Table X: Effect of Succinamic Acid-Cellulose Ratio in Amic Acid-Wood Pulp Reaction

Moles Succinamic Acid/AHGU	Reactant Conc. %	PUR	Total Ester Meq/AHGU	Half-Acid Ester Meq/AHGU	Crosslinks Meq/AHGU
6.98	72	7.0	0.161	0.122	0.020
5.23	54	7.0	0.202	0.151	0.025
3.51	54	4.7	0.249	0.201	0.024
3.49	36	7.0	0.244	0.185	0.029

Reaction Conditions: Aqueous solution, 3% ammonium sulfamate on weight of succinamic acid, wet formed wood pulp mat, cure 150°C 22 minutes.

shows that the extent of reaction is doubled with two treatments when the fabric is washed to remove unreacted material between treatments. If the fabric is not washed between treatments, the extent of reaction is tripled, indicating that some of the reactant from the first treatment is still in its reactive form and has not decomposed to some inactive species such as succinimide. In this case 70% of the amic acid available to the cellulose has reacted.

Table XI: Effect of Repeat Treatments on Increasing Reaction in Succinamic Acid - Rayon Fabric Reaction

Conc. of Succinamic Acid in Bath - %	Number of Treatments	Total Ester Meq/AHGU	Half-Acid Ester Meq/AHGU	Crosslinks Meq/AHGU
9	1	0.140	0.102	0.019
9/9	2*	0.268	0.227	0.020
9/9	2	0.432	0.352	0.040
18	1	0.157	0.123	0.018
18/18	2	0.366	0.332	0.017
18/18/18	3	0.486	0.429	0.028

Reaction Conditions: 2.0 PUR, cure 150°C 6 minutes, 1.08% sulfamic acid catalyst on weight of bath.
*After first treatment rinse in water until color is gone, then rinse in acetone, rinse in 5-10% acetic acid, rinse in water 3 times, rinse in acetone and air dry from acetone. Each rinse is approximately 5 minutes.

If the concentration of succinamic acid in the treating solution is increased, the amount of reaction is increased, but the efficiency of reaction is greatly reduced. In two treatments the extent of reaction is the same as with the lower amic acid concentration, although the cure time is the same.

If the succinamic acid concentration and cure time are increased greatly, D.S. values of 1.5 for total ester and 1.0 for half-acid ester are obtained (Table XII.). These values are quite high for a heterogeneous pad-bake system.

With the achievement of these relatively high D.S. values, it was of interest to see if other substrates would behave similarly. Cotton has been shown to give less reaction than rayon (2). In Table XIII it is seen that structure apparently limits the extent of reaction with cotton more than with rayon. At the time it was thought that structure limitations could be overcome by using never dried cotton but recent studies on the structure of never

Table XII: Repeat Treatments with Higher Concentration of Succinamic Acid in Cellulose-Amic Acid Reaction

Number of Treatments	Total Ester Meq./AHGU	Half-Acid Ester Meq./AHGU	Crosslinks Meq./AHGU
1	0.590	0.468	0.061
2	0.927	0.702	0.112
3	1.294	0.833	0.230
4	1.571	0.986	0.298

Reaction Conditions: 36% aqueous solution of succinamic acid plus 1.08% sulfamic acid, 4.0 PUR, cure 140°C 28 minutes, viscose rayon fabric, no wash between treatments.

dried cotton fibers indicate that the small increase in reaction over dried cotton fibers is not so unexpected (15, 16,17). The reaction with wood pulp falls between cotton and rayon in a manner similar to the openess of structure or the average disordered fraction measured by various techniques (18). The result of the saponified cellulose acetate fibers is somewhat consistent with the structure comparison although D.S. values in the range of those with rayon could be expected. The low reaction with polyvinylalcohol fibers, together with data on reaction of amic acids with simple alcohols (19), indicate that reaction with secondary alcohols is somewhat slower.

In Table XIV, never dried rayon, which is more accessible than dried rayon, does give more reaction than dried rayon.

The structure concept is explored further in Table XV where cotton is decrystallized. When water is used to remove the decrystallizing agent the extent of reaction is increased greatly. Some recrystallization of the cellulose chains does take place in the presence of water (20), possibly decreasing the amount of reaction. If solvents of lower hydrogen bonding capacity, less polarity and more bulk are used to remove the sodium hydroxide, less recrystallization and more reaction should take place, as is the case with dimethylformamide and isopropyl alcohol (20,21). However, structure does not appear to be the only limiting factor since the amount of reaction with viscose rayon is not increased by the sodium hydroxide treatment.

Maleamic and phthalamic acids were shown to be closest in reactivity to succinamic acid in earlier studies (5). In Table XVI the application of the high ratio of amic acid to cellulose, longer cure time, and repeat treatment procedures were applied to these acids with the unexpected result that the D.S. values did not increase in the same manner as with succinamic acid. The

Table XIII: Effect of Substrate in Cellulose - Amic Acid Reaction

Substrate	Total Ester Meq./AHGU	Half-Acid Ester Meq./AHGU	Crosslinks Meq./AHGU
Woven Viscose Rayon Fabric	0.855	0.594	0.131
Woven Cotton Fabric	0.178	0.165	0.006
Never Dried Cotton Fibers Wet Formed Mat	0.176	0.212	0.000
Wood Pulp Wet Formed Mat	0.249	0.201	0.024
Saponified Cellulose Acetate Fibers Kept Wet Wet Formed Mat	0.387	0.401	0.000
Crosslinked Polyvinyl Alcohol Fibers Wet Formed Mat	0.085	0.054	0.015

Reaction Conditions: 54% aqueous succinamic acid solution plus 1.62% ammonium sulfamate, 4.0 PUR, cure 150°C 22 minutes.

increase in crosslinking with no increase in half-acid ester content in the repeat treatment with maleamic acid is in definite contrast to the results in Table XI. No satisfactory explanation of these results can be made until the mechanism of reaction is known. Studies in this area are in progress (19).

In a further effort to understand the reaction involved, various specialized amic acids were looked at (Table XVII). The amine group in asparagine suppresses the reaction, indicating that the acid part of the molecule takes part in the reaction. Participation of the amide group is also indicated by the results of hydantoic acid, where again no reaction took place. An attempt was made to tie up the amine group in asparagine with other acids. Acetic acid gave no increase in reaction. Phosphoric acid in equimolar quantities with asparagine gave some reaction but direct esterification by the phosphoric acid was a contributing factor.

If the amide part of the molecule is reacting, then substituted amides should give less reaction, depending on the

Table XIV: Effect of Never Dried Rayon in Amic Acid - Cellulose Reaction

Type of Rayon	Total Ester Meq/AHGU	Half-Acid Ester Meq/AHGU	Apparent X-links Meq/AHGU
Woven Fabric	0.297	0.264	0.016
Never Dried Fibers Wet Formed Mat	0.355	0.303	0.026
Dried Never Dried Fibers	0.291	0.250	0.020

Reaction Conditions: 18% aqueous succinamic plus 1.08% sulfamic acid, 4.0 PUR, cure 150°C 9 minutes.

Table XV: Effect of Decrystallization of Substrate on Succinamic Acid - Cellulose Reaction

Decrystallization Treatment	Rinse Solvent	Total Ester Meq/AHGU	Half-Acid Ester Meq/AHGU	Apparent Crosslinks Meq/AHGU
A. Cotton Fabric				
None	None	0.178	0.164	0.006
17.5% NaOH, -7.5°C 3 hrs.	Water	0.349	0.291	0.029
17.5% NaOH, -7.5°C 3 hrs.	DMF	0.382	0.330	0.026
17.5% NaOH, -7.5°C 3 hrs.	Isopropyl Alcohol	0.502	0.426	0.038
B. Viscose Rayon Fabric				
None	None	0.855	0.594	0.131
17.5% NaOH, -7.5°C 3 hrs.	Isopropyl Alcohol	0.861	0.610	0.125

Reaction Conditions: 54% succinamic acid, 1.6% ammonium sulfamate, 4.0 PUR, cure 150°C 22 minutes.

substituent. If the acid groups is reacting, then all the ester should be half-amide ester. The last three results in Table XVII show again that there is no clear distinction, and that both groups can be contributing.

The theme of this paper has been the unique and surprising aspects of the amic acid-cellulose reaction. Several of the

Table XVI: Effect of Reactant Type in Cellulose - Amic Acid Reaction and Phthalamic Acids at High Concentrations

Amic Acid	Number of Treatments	Total Ester Meq/AHGU	Half-Acid Ester Meq/AHGU	Apparent Crosslinks Meq/AHGU
Succinamic	1	0.590	0.468	0.061
Succinamic	2	0.927	0.702	0.112
Maleamic	1	0.245	0.242	0.002
Maleamic	2	0.326	0.240	0.043
Phthalamic	1	0.129	0.127	0.001

Reaction Conditions: 0.46 molar aqueous amic acid except for phthalamic acid limited to 0.31 molar for solubility reasons, plus 6% on weight of amic acid ammonium sulfamate, viscose rayon fabric, 4.0 PUR, cure 150°C 22 minutes.

Table XVII: Effect of Reactant Type in Cellulose-Amic Acid Reaction

Amic Acid	Total Ester Meq/AHGU	Half-Acid Ester Meq/AHGU	Amide Ester Meq/AHGU	Crosslinks Meq/AHGU
Succinamic Acid	0.231	0.195	-	0.018
L- Asparagine $HOC(O)CH(NH_2)CH_2C(O)NH_2$	0.000	0.000	-	-
Hydantoic Acid $HOC(O)CH_2NHC(O)NH_2$	0.000	0.000	-	-
Succinic Anhydride plus Ethanolamine	0.121	0.066	-	0.055
Succinic Anhydride plus Glycine	0.050	0.025	0.002	0.012
Succinic Anhydride plus Ethylenediamine	0.016	0.000	0.004	0.006

Reaction Conditions: 0.154 molar solution of amic acid, 2.0 PUR, viscose rayon fabric, cure 150°C 6 minutes .

Table XVIII: Effect of Reactant Type in Amic Acid - Cellulose Reaction Exploration of Succinic Acid Reaction

Reactant	Reaction Conditions	Total Ester Meq/AHGU	Half-Acid Ester Meq/AHGU	Crosslinks Meq/AHGU
Succinic Acid	A	0.038	0.014	0.012
Succinamic Acid	A	0.184	0.165	0.009
Succinamide	B	0.042	0.011	0.016
Succinic Acid plus Succinamide	D	0.115	0.061	0.027
Succinamic Acid plus Succinamide	D	0.165	0.119	0.023
Succinic Acid plus Phthalamide	D	0.074	0.077	0.002
Succinic Acid plus Acetamide	D	0.054	0.051	0.000
Succinic Acid	C	--	0.079	--
Succinamic Acid	C	--	0.120	--
Succinic Acid	E	0.541	0.439	0.051
Succinamic Acid	E	0.725	0.536	0.095

Reaction Conditions: A - 0.154 molar aqueous solution, 2.0 PUR, cure 150°C 6 minutes
B - 0.072 molar aqueous solution, 2.0 PUR, cure 150°C 6 minutes
C - 0.072 molar formamide solution, 2.0 PUR, cure 150°C 8 minutes
D - 0.072 molar aqueous solution of each reactant, 1.08% ammonium sulfamate, cure 150°C 6 minutes
E - 0.46 molar aqueous solution plus 7.8 % NH_3 with succinic acid, 1.62% ammonium sulfamate, cure 150°C 22 minutes.

tables presented here have given results indicating that the acid and amide groups may act together to give reaction with cellulose that is not realized by succinic acid and succinamide, molecules containing either the acid or the amide groups alone (Table XVIII, 5). The results in Table XVIII show the last surprising aspect of the amic acid reaction .- the formation of cellulose half-acid-acid esters by combination of acid and amide groups not on the same molecule. Although the extent of reaction is not as great as with succinamic acid, the combination of succinic acid with such diverse amides as succinamide, acetamide, or phthalamide give significantly more reaction than with succinic acid alone. This same trend is seen when formamide is used as solvent. Finally, by use of the increased ratio of reactant to cellulose with sulfamic acid as catalyst, a relatively high amount of reaction is obtained with the ammonium salt of succinic acid. The fabric was noticably tendered, however. Such tendering does not occur with the amic acids (5).

Conclusions

By a careful balance of amic acid type, accessibility of the cellulose substrate, amic acid-cellulose ratio, and cure time and temperature, it was found that the amic acid-cellulose reaction can be utilized to produce essentially all cellulose half-acid ester at a D.S. value below about 0.3. Substitution values significantly above 0.3 can be realized, but some crosslinking occurs, the amount depending on the severity of the reaction conditions. Several unexplained results were obtained in this study, indicating that some aspects of the amic acid reaction are unique. Although the object of this work was not to define the mechanism of reaction, several results led to a belief that the acid and the amide groups exhibit a synergestic effect, probably in a stereospecific manner. A working model has been developed to explain the unusual results obtained here.

$$HOC(O)RC(O)NH_2 \rightleftharpoons {}^{-}OC(O)R\overset{OH}{\underset{+}{C}}-NH_2$$

$$\longrightarrow \text{Cellulose} - CH_2\underset{H}{O}-\overset{{}^{-}O-C(O)-R}{\underset{OH}{C}}-NH_2$$

$$\longrightarrow \text{Cellulose} - CH_2OC(O)-R-\overset{NH_2}{\underset{OH}{C}}-OH$$

$$\longrightarrow \text{Cellulose}-CH_2OC(O)-R-CO_2H + NH_3$$

The empirical information obtained in this study was then used to prepare cellulose semi-esters of varying D.S. and crosslinking values for evaluation of selected properties listed earlier. Detailed results of these studies will be reported later.

Acknowledgements

We wish to thank Cotton, Inc. and American Enka Co. for their support of parts of this work. We are also thankful to Dr. Frank Cramer for his helpful advice on the pendant carboxyl test method, particularly in regard to the correction factor. Special thanks also go to the many individuals who contributed to this effort.

Literature Cited

1. Allen, Thomas C. and Cuculo, John A, J. Polymer Sci.: Macromolecular Reviews, (1973),7, 189-262.
2. Cuculo, J. A., Textile Res. J.,(1971), 41 (4),321-326.
3. Cuculo, J. A., U.S. Patent 3,555,585 (1971).
4. Cuculo, J.A., Textile Res. J.,(1971), 41 (5), 375-378.
5. Johnson, Emmitt H. and Cuculo, John A., Textile Res. J.(1973) 43 (5), 283-293.
6. Johnson, Emmitt H., Thesis for Master of Science Degree(1972) Department of Textile Chemistry, North Carolina State University, Raleigh, N. C.
7. Cramer, Frank B., E.I. DuPont de Nemours and Co., Private Communication, (1972).
8. Genung, L. B. and Mallatt, R. C., Ind. Eng. Chem., (1941), 13, 369-374.
9. Allen, T. C., Textile Res. J., (1964), 34, 331-336.
10. Bender, M. L., Chloupek, F., and Neveu, M. C., J. Amer. Chem. Soc., (1958), 80, 5384-5387.
11. Malm, C. J. and Fordyce, C. R., Ind. Eng. Chem., (1940), 32, (3), 405-408.
12. Rowland, S.P., Welch, C. M., Brannan, M. A. F., and Gallagher, D. M., Textile Res. J., (1967), 37, 933-941.
13. Rowland, S.P., Welch, C. M., and Brannan, M. A. F., U. S. Patent 3,526,048 (1970).
14. Bullock, A. L., Vail, S. L., and Mack, C. H., U. S. Patent 3,294,779 (1966).
15. Ingram, P., Wood, D. K., Peterlin, A., and Williams, J. L., Textile Res. J., (1974), 44 (2), 96-106.
16. Williams, J. L., Ingram, P., Peterlin, A., and Woods, D. K., Textile Res. J., (1974), 44 (5), 370-377.
17. Morosoff, N., J. Appl. Polymer Sci., (1974), 18 (6),1837-1854.
18. Tripp, V. W. in N. M. Bikales and L. Segal, Eds., "Cellulose Derivatives", Vol. V in High Polymers, Part 5, p.307, Wiley-Interscience, New York.
19. Bowman, B. G., Unpublished Work (1974).
20. Jeffries, R. and Warwicker, J. O., Textile Res. J., (1969), 39 (6), 548-559.
21. Warwicker, J. O., Jeffries, R., Colbran, R. L., and Robinson, R. N., "A Review of the Literature on the Effect of Caustic Soda and Other Swelling Agents on the Fine Structure of Cotton", p. 133, Shirley Institute Pamphlet No. 93, The Cotton Silk and Man-Made Fibres Research Association, Shirley Institute, Didsbury, Manchester, England,(1966).

Work performed in Fiber and Polymer Science Program, School of Textiles, North Carolina State University

6

Wood and Wood-based Residues in Animal Feeds

ANDREW J. BAKER and MERRILL A. MILLETT
Forest Products Laboratory, Madison, Wis. 53705[1]
LARRY D. SATTER
Department of Dairy Science, University of Wisconsin, Madison, Wis. 53706

Cellulose is the most abundant, naturally renewable material on earth. It, and hemicellulose, make up about 70% of the dry weight of shrubs and trees. The cellulose of woody plants, however, is largely unavailable to ruminants because of the highly crystalline nature of the cellulose molecule and the existence of a lignin-carbohydrate complex. If convenient ways can be found to enhance the availability of wood cellulose to enzymatic or microbiological systems, wood residues could provide an additional renewable energy feed supply for a world that can maintain no contingency reserve of feedstuffs. It would permit utilization of the large quantities of cellulosic residues that occur during harvest and manufacture of wood and cellulose products and provide a method of disposal of the used products.

This article presents a summary of research conducted on the use of wood and wood-based materials in animal feeds at the Forest Products Laboratory and the University of Wisconsin, and research in cooperation with the Tennessee Valley Authority, the U.S.D.A. Agricultural Research Service, Animal Nutrition Laboratory, Pennsylvania State University, and Auburn University.

Animal Feeding Studies

Early Research. Efforts by the Forest Products Laboratory to utilize wood in animal feeds began in 1920 when eastern white pine and Douglas-fir sawdust were hydrolyzed and fed to animals at the University of Wisconsin and the U.S. Department of Agriculture, Beltsville, Md. The work was started as a result of high feed grain prices during 1918-19. Wood was hydrolyzed and the washings and hydrolyzate were neutralized, concentrated, mixed with the unhydrolyzed residue and dried (1).

[1] Forest Service, U.S. Department of Agriculture. Maintained at Madison, Wis., in cooperation with the University of Wisconsin.

This type of material was used in several feeding experiments with sheep and dairy cows (2-4). Results indicated that certain animals could eat rations containing up to one-third hydrolyzed sawdust mixture. Animals requiring considerable energy intake such as dairy cows could eat up to 15% of the hydrolyzed mixture without noticeable milk production effects. It was determined that the eastern white pine mixture was 46% digestible and that the Douglas-fir mixture was 33% digestible. It was concluded that feeding hydrolyzed wood was practical only when natural feed grains were in short supply.

Research on wood hydrolysis was conducted in the 1940's to produce concentrated sugar solutions suitable for stock and poultry feed. Over 200 tons of molasses were produced in pilot plants and sent to universities, agricultural experiment stations, and other agencies for feeding tests with milk cows, beef cattle, calves, lambs, pigs, and poultry (5,6). In general, the tests indicated that wood-sugar molasses is a highly digestible carbohydrate feed comparable to blackstrap molasses. In addition, the protein value of torula yeast, grown on neutralized dilute wood hydrolyzate, was found to be equivalent to casein when supplemented with methionine (7). Torula yeast has also been produced in three North American plants on the residual sugars in spent sulfite pulping liquors. Two plants are now operating.

Results from feeding tests with wood molasses led to production during the early 1960's of a concentrated hemicellulose extract called Masonex, a byproduct from hardboard production by the Masonite Corporation (8).

Current Studies. Recent research on the use of wood and wood residue in animal feeds was started as one approach to utilize the vast quantities of residue from logging, lumber and plywood manufacturing, and pulp and papermaking. Wood residue may serve as a source of digestible energy or as a roughage in ruminant rations. Fattening feedlot cattle, as well as lactating dairy cattle, need a minimum of fibrous feed in their ration and it is conceivable that indigestible fibrous wood residues could play a non-nutritive role in ruminant nutrition. It has been estimated that all of the wood and bark residues would supply more than enough roughage for all concentrates fed in the United States (9). In addition, more than 1.7 million tons of partially digestible pulp and papermaking fiber residues are produced annually that could supplement feed grains as sources of energy.

Animal feeding studies were conducted to determine acceptability, palatability, and digestibility of wood and bark residues to determine their value as roughage substitutes. Various physical and chemical methods to increase cellulose availability to rumen micro-organisms were evaluated with *in vitro* rumen methods. Digestibility trials were then conducted to determine the in vivo digestibility of products from selected treatments. Pulp and papermill fiber residues were also evaluated by chemical analysis, in vitro and in vivo methods. Rations containing as much as 80% fiber residues were fed to animals through a complete reproductive cycle to determine long-term effects on general health and reproductive capacity.

In Vitro Assay Methods. The dry matter digestibility of various wood species and of the effects of chemical and physical pretreatments on digestibility was determined by the in vitro rumen method of Mellenberger, et al. (11). Results are reported as percent weight loss after 5 days of incubation at 39° C. An enzyme method was developed to provide an alternative assay procedure that did not depend upon the availability of a rumen fistulated cow (12). This method utilizes "Onazuka" SS enzyme obtained from Trichoderma veride in an acetate buffer and usually a 10-day incubation period. Digestibility is determined by analyzing the solution before and after incubation to determine the increase in reducing substances. The results of this test do not directly indicate rumen digestibility but they do indicate changes in digestibility.

The in vitro rumen test indicated that the digestibility of all wood species is low (13,11). All softwoods or coniferous species are essentially nondigestible. Hardwoods, or deciduous species, are somewhat digestible. Digestibility of the wood and bark of several tree species is shown in Table I. Note that the digestibility of soft maple wood is about 20%, aspen wood is about 33%, and aspen bark is about 50%.

Figure 1 shows results of feeding trials with red oak wood (14) and aspen wood and bark (15) and a method for estimating the in vivo digestibility by extrapolation of the data to 100% wood or bark. The red oak trial was with sheep, and the aspen wood and bark trial was with goats. Thus for red oak, the estimated in vivo digestibility is 0%; for aspen wood it is estimated to be about 40%, and for aspen bark it is about 50%. This indicates that aspen wood and bark could supply considerable digestible energy as well as roughage for ruminants.

Wood Residues as an Alternate Source of Roughage. Even though most untreated woods can contribute little to the dietary energy needs of ruminants, wood can still serve a useful function as a roughage substitute. Roughage is required in the ration to provide tactile stimulation of the rumen walls and to promote cud-chewing, which in turn increases salivation and supply of buffer for maintenance of rumen pH. Roughage materials currently used include hay, corn cobs, cottonseed hulls, oat hulls, rice hulls, and polyethylene pellets. A roughage substitute should be: readily obtained at low cost, effective at low levels, uniform in chemical and physical characteristics, capable of easy and uniform

Table I

in vitro Dry-Matter Digestibility of Various Woods and Their Barks

Substrate	Digestibility[a] Wood	Bark	Substrate	Digestibility[a] Wood	Bark
	%	%		%	%
HARDWOODS			HARDWOODS--continued		
Red alder	2	--	Soft maple small twigs	37	--
Trembling aspen	33	50	Sugar maple	7	14
Trembling aspen (groundwood fiber)	37	--	Red oak	3	--
Bigtooth aspen	31	--	White oak	4	--
Black ash	17	45	SOFTWOODS		
American basswood	5	25			
Yellow birch	6	16	Douglas-fir	5	--
White birch	8	--	Western hemlock	0	--
Eastern cottonwood	4	--	Western larch	3	7
American elm	8	27	Lodgepole pine	0	--
Sweetgum	2	--	Ponderosa pine	4	--
Shagbark hickory	5	--	Slash pine	0	--
Soft maple	20	--	Redwood	3	--
Soft maple buds	36	--	Sitka spruce	1	--
			White spruce	0	--

[a]For comparison: Digestibility of cotton linters was 90%; of alfalfa, 61%.

mixing, maintain normal rumen functions and feed intake, and able to prevent rumen parakeratosis and liver abscesses (16). If it is used in dairy rations, it should maintain normal milk fat test.

The roughage qualities of red oak sawdust have been determined by feeding beef cattle and sheep (17-19). In addition to the usual criteria of weight gain and efficiency of feed conversion, such measurements of carcass quality as grade, rib-eye area, and fat marbling were also noted. Attention was focused on livers and stomachs at slaughter, because abnormalities in these organs are characteristic of animals on roughage-deficient diets. It was concluded that oak sawdust was an effective roughage substitute when used as 5 to 15% of the total ration.

Roughage is necessary in dairy cow rations to prevent abnormally low milk fat tests (20). For economic reasons it is desirable to produce milk of high fat content. Hay supplies are, at times, limited and costly in some areas. In these areas it would be desirable to have an alternate roughage that would meet the "roughage requirement" for lactating dairy cows, that is not seasonal and would be compatible with automated feeding systems. Aspen sawdust, which is about 35% digestible, was fed at various concentrations to lactating dairy cows to determine if part or all of the hay could be replaced when feeding high-grain rations.

One feeding experiment (21) with lactating cows shows that aspen sawdust was effective as a partial roughage substitute in a high-grain dairy ration. The aspen sawdust was air-dried and hammermilled to pass through a screen plate with 1/8-inch-diameter holes. Cows maintained a normal milk fat level on 2.3 kg. of hay and about 17 kg. of pelleted grain, one-third of which was aspen sawdust. Cows receiving a similar ration without sawdust had a milk fat content half as great. The ratio of ruminal acetate to propionate was much higher in the cows fed aspen. Inclusion of aspen in a high-concentrate ration nearly doubled ruminating time. If less dietary aspen would be equally as effective in complete pelleted dairy rations, aspen sawdust could become an attractive roughage substitute in areas where hay is expensive and difficult to obtain.

In a second experiment (22), combining various levels of aspen sawdust with 5% bentonite and 2% sodium bicarbonate (based on the total ration), it was found that aspen sawdust could be a roughage extender or a partial roughage substitute in high-concentrate dairy rations. Sawdust maintained fat test and diminished off-feed problems when constituting about 30% of the ration dry matter in high or all-concentrate dairy rations. Since the dry matter digestibility of aspen sawdust was less than for other ration components, cows eating sawdust-containing rations compensated for the lower digestibility by eating more of the ration; thus, cows maintained total digestible energy intake. Whether high-producing cows already at maximum feed intake could do this is questionable.

Aspen sawdust has useful roughage characteristics, but using it as the only roughage in high-concentrate dairy rations cannot be recommended. Approximately 30% of the ration dry matter would have to be sawdust; that is too high to be practical because the cows would have trouble consuming that large a volume of feed. Sodium bentonite and sodium bicarbonate apparently have an additive effect toward maintaining fat test when combined with aspen sawdust. In combination with bentonite and bicarbonate, smaller quantities of sawdust would probably be sufficient to maintain a given fat content of milk.

As little as 2.3 kg. of hay/cow per day is effective in stabilizing feed intake. To supplement the hay, adding 10-15% of the high-concentrate diet as aspen sawdust, 5% as sodium bentonite, and 2% as sodium bicarbonate might extend limited forage supplies. Since aspen sawdust does not serve well as the sole source of roughage in a complete all-concentrate ration, its potential appeal as a forage substitute for lactating dairy cows is reduced.

Pretreatments to Increase Digestibility

Several physical and chemical pretreatments were tested for their ability to increase digestibility of wood cellulose. The treatments were electron irradiation, vibratory ball milling, gaseous and liquid ammonia, gaseous sulfur dioxide, dilute sodium hydroxide, and white-rot fungi (23-25). Each of the treatments is capable of producing a product at high yield without a waste stream or byproduct.

The digestibility response to the various treatment conditions was followed by *in vitro* rumen and cellulase digestion assay procedures. Larger quantities of products of selected treatments were prepared for animal digestion trials with goats to determine *in vivo* digestibility and to observe palatability and acceptability. Goats were selected because they are small ruminants and require less space and feed.

High-Energy Electron Irradiation. The effect of exposure to increasing levels of electron irradiation on the *in vitro* digestibility of aspen and spruce is shown in Table II. Aspen carbohydrate digestion is essentially complete if it is assumed that only carbohydrate has been solubilized at an electron dosage of 10^8 rep. (roentgen equivalent physical). However, the lignin content of this aspen was 19.5%, and it might be expected that some lignin degradation products would be formed at this dosage level. If water soluble, these would contribute to the figure for dry matter digestibility. In any event, electron irradiation is an effective means for enhancing the digestibility

Table II

Effect of Electron Irradiation on *in vitro* Rumen Digestibility of Aspen and Spruce

Radiation dosage	Digestibility	
	Aspen[a]	Spruce
rep.[b]	%	%
0	55	3
10^6	52	3
10^7	59	5
5×10^7	70	8
10^8	78	14

[a] *Populus tremuloides*. This sample was from a board containing a high proportion of tension wood fibers. Tension wood is characterized by an exceptionally high carbohydrate-to-lignin ratio; thus, the high digestibility of this untreated aspen sample in comparison with that shown in Table I.

[b] Roentgen equivalent physical.

of aspen. It does very little to improve digestibility of spruce, however; the maximum digestibility was only 14% at the highest dosage level. Although higher dosage levels would probably improve digestibility further, they would also increase the level of carbohydrate destruction. From earlier work on the use of electron irradiation to enhance wood saccharification (26) it was shown that carbohydrate destruction was about 15% at 10^8 rep. and increased to about 45% at 5×10^8 rep. The product of the latter dosage was almost completely water soluble and was strongly acidic.

Vibratory Ball Milling. The effect of vibratory ball milling on the *in vitro* rumen digestibility of aspen and red oak is shown graphically in Figure 2. *In vitro* digestibilities of both woods increased rapidly with milling time to about 30 min. and then increased more slowly with further milling. Digestibility was highly dependent on time of *in vitro* rumen incubation; at least 5 days of incubation were required for digestibilities to attain 90% or more of their plateau values.

In vitro rumen digestibility of aspen and red oak which had been milled for 240 min. was 80% and 67%, respectively. Results of an enzymatic hydrolysis of the milled products using a cellulase demonstrated that this was not merely a solubilization effect. The 240-min. milled aspen and oak produced 63% and 57%, respectively, of their weight as glucose after enzyme digestion. Sugar production from the unmilled aspen and oak was 10.0% and 0.0%, respectively. Of the total carbohydrates in aspen and red oak, 70-80% was made accessible to cellulase digestion by vibratory ball milling.

In Figure 3 *in vitro* rumen digestibility is plotted as a function of milling time for five hardwood species. The digestibility values are those obtained with 5-day incubation. The first 20-30 min. of milling appear to have the major influence on digestibility. A digestibility plateau is apparently attained beyond which additional milling is of little value.

It is difficult to ascribe definite reasons for the wide variation in response between the woods. Certainly particle size alone is not the governing factor. All wood samples received the same degree of milling, and settling tests in water indicated similar particle size distribution. The controlling factor must be the quantity, chemical nature, and distribution of lignin.

The very selective response of the various species to vibratory ball milling makes this technique of limited value as a general means for upgrading the digestibility of wood residues. Moreover, there is a question whether finely ground wood will function as effectively in the ruminant as it does in *in vitro* assay. With forages, fine grinding has increased the *in vitro* digestibility of cellulose, but it has not produced similar

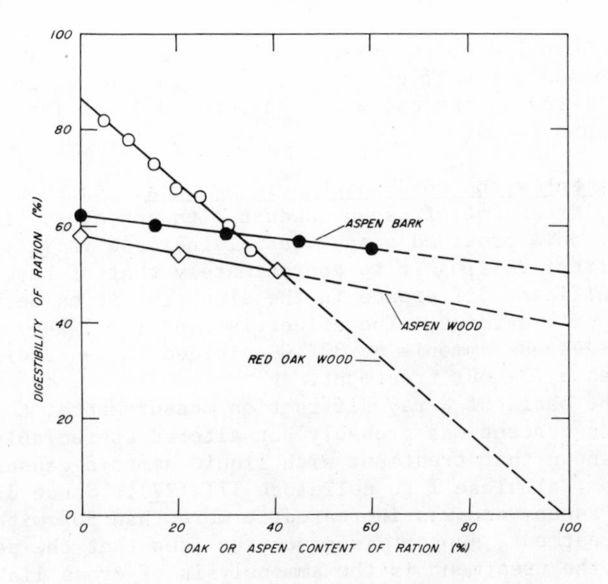

Figure 1. In vivo *digestibility of red oak and aspen wood and aspen bark*

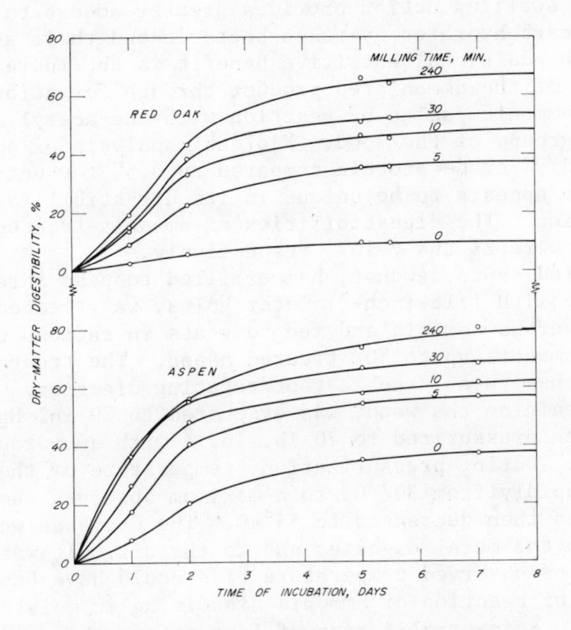

Figure 2. Relation of in vitro *rumen digestibility of red oak and aspen to time of* in vitro *rumen incubation and extent of vibratory ball milling*

responses when fed to ruminants, when digestibility in fact has been decreased. Insufficient residence time in the rumen has been postulated as the cause of the lowered digestibility of finely ground feeds.

Treatment with Anhydrous Liquid Ammonia. As shown in Table III, treatment of aspen sawdust with anhydrous liquid or gaseous ammonia provided a substantial increase in *in vitro* digestibility, raising it to approximately that of hay. There is no significant difference in the digestibilities between the two types of treatment. The effect is rapid; a 1/2-hour treatment with gaseous ammonia at 30° C. yielded the same digestibility value as a 73-hour treatment.

On the basis of X-ray diffraction measurements, total crystalline content was probably not altered sppreciably, but it has been shown that treatment with liquid ammonia causes a phase change from cellulose I to cellulose III (27). Since digestibility of aspen wood was increased to more than 50% with liquid ammonia treatment, support is given the idea that the pertinent action of the treatment is the ammonolysis of cross links of glucuronic acid esters (28).

Hardwoods which have been treated with liquid ammonia and air dried have a markedly increased swelling capacity in water (29). This swelling action provides greater access to the structural carbohydrates by rumen bacteria and their associated enzymes. An additional nutritive benefit is the increased nitrogen content of the ammoniated product through formation of amides and ammonium salts by reaction with the acetyl and uronic acid ester groups of the wood. Kjeldahl analysis of ammoniated aspen showed 9% crude protein compared to 0.5% for untreated wood. Aspen appears to be unique in its digestibility response to ammoniation. The digestibilities of ammonia-treated spruce and red oak were 2% and 7-10%, respectively.

Air-dried aspen sawdust, hammermilled to pass through a screen plate with 1/16-inch-diameter holes, was treated with gaseous anhydrous ammonia and fed to goats in rations containing increasing amounts up to 50% treated aspen. The treatment was done in batches in a 13-cubic-foot rotating digester. The digester, containing the wood, was evacuated to 20 in. Hg for 20 min. and then pressurized to 70 lb. in.$^{-2}$ with anhydrous ammonia for 2 hours. During pressurization, temperature of the wood increased rapidly from 30° C. to a maximum observed temperature of 74° C. and then decreased to 55° C. The decrease was due to heat loss to the metal digester and to the air. It was calculated that the observed temperature rise could have been caused by the heat of reaction of ammonia dissolving in moisture present in the wood. No neutralization of free or adsorbed ammonia on the product was attempted. Ammonia smell from the product was not noticeable after airing the product on the floor for 1 week.

Table III

in vitro Rumen Digestibility of Aspen Sawdust Exposed to Anhydrous Liquid and Gaseous Ammonia

Treatment[a]		
Chemical	Time	Digestibility
	hr	%
Control	--	33
Liquid NH_3	1	51
Gaseous NH_3	1/2	48
	2-1/2	47
	16	46
	73	46

[a]At 30° C.

Table IV

Effect of Alkali Treatment on the *in vitro* Rumen Digestibility of Various Hardwoods

Species	Yield	Control	Treated[a]
	%	%	%
Trembling aspen	87	33	55
Bigtooth aspen	90	31	49
Black ash	91	17	36
American basswood	89	5	55
White birch	92	8	38
Yellow birch	94	6	19
Eastern cottonwood	93	4	11
American elm	93	8	14
Soft maple	92	20	41
Red oak	94	3	14
White oak	90	4	20

[a] 5-g wood treated for 1 hr with 100 ml of 1% NaOH, washed to neutrality, and dried.

A digestion trial with goats, as was done with aspen bark, indicated an extrapolated *in vivo* dry-matter digestibility of 50%.

Treatment with Aqueous Sodium Hydroxide. The results of *in vitro* rumen digestion show a range of response to the alkali treatment for the various species investigated (Table IV). Aspen and basswood, attaining a digestibility of 55%, are outstanding in their response to alkali pretreatment. The tenfold increase for basswood is especially intriguing. Bigtooth aspen is only slightly less digestible than trembling aspen. Black ash, white birch, and soft maple show an intermediate response with digestibilities ranging between 35% and 40%. The other species have digestibilities of less than 20%. Douglas-fir and Sitka spruce, which are softwoods with a maximum *in vitro* digestibility of 1% and 2%, respectively, did not respond to the alkali treatment. The difference in response appears to be related to the lignin content of the treated hardwoods (Figure 4).

To better define conditions for optimum processing, aspen was treated at room temperature with 0.5% and 1.0% solutions of sodium hydroxide at various liquid-to-solid ratios. Then it was washed to neutrality, dried and assayed. The results in Table V show that from 5-6 g. of NaOH per 100 g. of wood are necessary for a maximum effect on *in vitro* digestibility. This was attained with a 12:1 liquor-to-wood ratio at the 0.5% alkali level or a 6:1 ratio with 1% alkali. It is interesting that the minimum quantity of sodium hydroxide needed for attaining maximum digestibility is roughly equivalent stoichiometrically to the combined acetyl and carboxyl content of the aspen. The main consequence of alkali treatment thus appears to be the breaking (by saponification) of intermolecular ester bonds (28,30). Rupture of these cross links promotes the swelling of wood in water beyond normal water-swollen dimensions; thus it favors increased enzymatic and microbiological penetration into the fine structure of wood. At optimum conditions (6 g. NaOH to 100 g. wood) the yield is about 95%. The 5% loss in weight is caused by saponification and removal of acetyl groups during the water wash.

Treatment with Sulfur Dioxide. It was found that gaseous sulfur dioxide can disrupt the lignin-carbohydrate association *in situ* and yield a product of high digestibility without physical removal of the lignin. Wood in the form of sawdust was reacted for 2 hours (hardwoods) or 3 hours (softwoods) at 120° C. with an initial SO_2 pressure at room temperature of 30 lb. in.$^{-2}$ and a water-to-wood ratio of 3:1 (no free liquid). After blowdown and a brief evacuation to remove adsorbed SO_2, the treated woods were neutralized to about pH 7 with sodium hydroxide and then air dried. Table VI presents analytical data and values for 48-hour cellulase digestion for both the original woods and

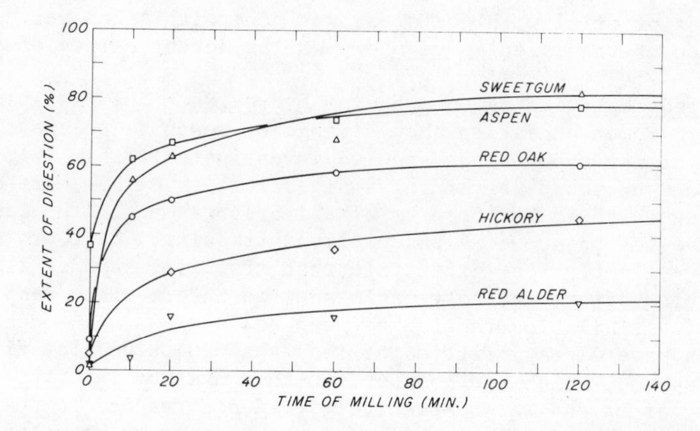

Figure 3. Relationship between in vitro *rumen digestibility and time of vibratory ball milling*

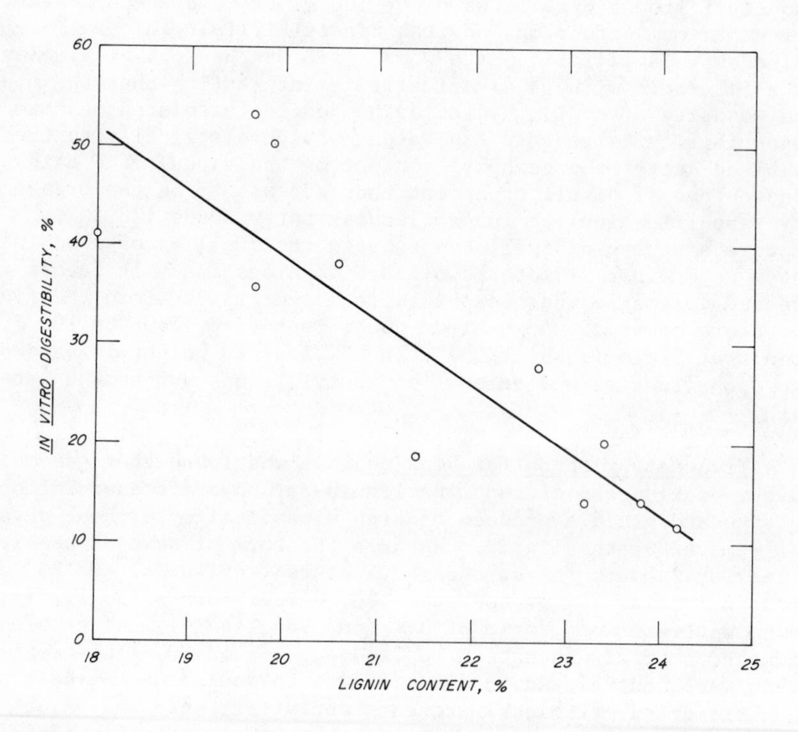

Figure 4. Relationship between lignin content and in vitro *digestibility for NaOH treated hardwoods* (30)

Table V

Effect of Alkali Treatment Variables on the *in vitro* Dry-Matter Digestibility of Aspen

NaOH concentration	Ratio of solution to wood	NaOH per 100 g wood	Treating time	Yield	Digestibility
%		gm	hr	%	%
0	0	0	0	100	37
0.5	4:1	2	2	98	47
	8:1	4	2	98	50
	12:1	6	2	95	55
	16:1	8	2	93	53
1.0	2:1	2	1	98	41
	4:1	4	1	96	48
	6:1	6	1	95	51
	8:1	8	1	94	50
	10:1	10	1	93	54
	20:1	20	1	87	50

Table VI

Composition and Cellulase Digestion of Various Woods

Before and After SO_2 Treatment

Species	Lignin		Carbohydrate		Digestibility	
	Before	After	Before	After	Before	After
	%					
Quaking aspen	20	7	70	71	9	63
Yellow birch	23	9	66	67	4	65
Sweetgum	20	5	66	64	2	67
Red oak	26	8	62	60	1	60
Douglas-fir	30	24	65	63	0	46
Ponderosa pine	31	19	59	58	0	50
Alfalfa	17	--	51	--	25	--

the treated products. Data for alfalfa is included for comparison.

Cellulase digestion of the original woods was minimal, from a high of 9% for aspen to essentially 0% for the two softwoods. Even with alfalfa, only half of the available carbohydrate was converted to sugars. Yields of the SO_2-treated products were 106-112% based on starting material, a result of the sulfonation and neutralization reactions. Although all of the lignin was retained in the products, Klason lignin analysis of the five treated hardwoods showed lignin values of only 5-9%. This suggested that the original lignin had been extensively depolymerized during SO_2 treatment and converted to soluble products, a fact subsequently confirmed by extraction with boiling water. Depolymerization was less extensive with the two softwoods, and the higher Klason values are reflected by a decreased digestibility. Enzymatic conversion of the hardwood carbohydrates was essentially quantitative, indicative of a complete disruption of the strong lignin-carbohydrate association in the original woods. The 60-65% digestibility of the treated hardwoods is comparable to the digestibility of a high quality hay. The two softwood products would be equivalent to a low quality hay, but might be upgraded through a better choice of processing conditions.

A 140-kg. batch of SO_2-treated material was prepared from red oak sawdust and fed to goats at levels of 0, 20, 35, and 50% of a pelleted forage ration over an 8-week period to obtain information on palatability, possible toxic side effects, and *in vivo* nutritional value. Average *in vivo* digestibilities for dry matter and carbohydrate as a function of wood content of the rations are plotted in Figure 5. Extrapolation of the curves to 100% SO_2-treated wood yielded values of about 52% for dry matter digestion and 60% for carbohydrate digestion. From the shallow slope of the curves, it appears that a vapor phase treatment with sulfur dioxide effectively converts red oak sawdust into a ruminant feedstuff having the digestible energy equivalence of a medium quality hay. Neutralization of the treated product with ammonia rather than sodium hydroxide would augment its protein equivalence.

Treatment with White-Rot Fungi. White-rot fungi decompose lignin as well as cellulose and hemicellulose in wood. Some remove lignin faster than they do the carbohydrates relative to the original percentage of each. The resulting decayed wood has a lower lignin content than that of the original wood.

Nine white-rot fungi were examined for their ability to remove lignin faster than polysaccharides from aspen and birch wood. During decay most of the fungi decreased the lignin content of the wood; that is, they removed a larger percentage of the lignin than of polysaccharides. Lignin removal was always accompanied by removal of polysaccharides. The decayed woods

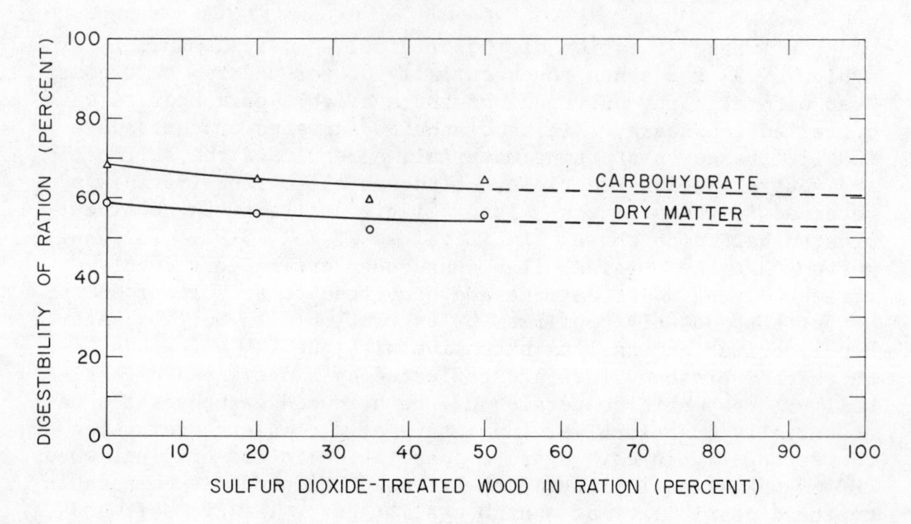

Figure 5. In vivo *dry-matter digestion of rations containing sulfur dioxide-treated red oak*

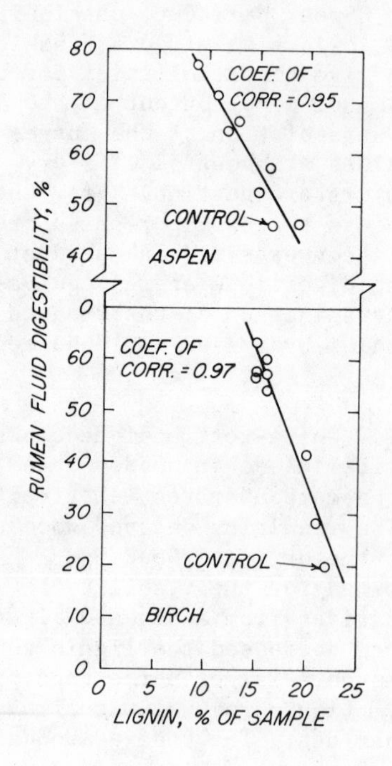

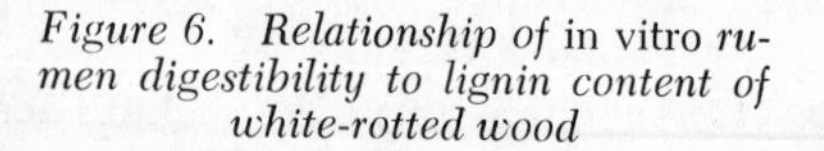

Figure 6. Relationship of in vitro *rumen digestibility to lignin content of white-rotted wood*

have higher in vitro rumen digestibility than the untreated wood and digestibility is inversely related to the lignin content as shown in Figure 6.

Pulp and Papermill Residues and Wood Pulp

Effect of Delignification on Digestibility. Lignin appears to be a major obstacle to microbiological attack of wood. Delignification would then seem to be a straightforward approach to making cellulose available to microbes. To obtain information on the effect of method and degree of lignin removal necessary to make various species digestible, a series of kraft pulps having a range of yields and lignin contents were prepared for in vitro rumen digestibility determination (31).

Four wood species were included: two hardwoods, paper birch and red oak; and two softwoods, red pine and Douglas-fir. Pulping variables were selected to produce pulps with yields from 40-80% and lignin content from 1-32%. Since hemicellulose is removed more rapidly than lignin during the early stages of pulping, some of the high-yield pulps have a higher percentage of lignin than the original wood.

Data showing the relationship between in vitro digestibility and extent of delignification for kraft pulps made from the four species are shown in Figure 7. Extent of delignification is the percent of the lignin removed from the original wood. It is calculated from pulp yield and lignin content of the original wood and the pulp.

Figure 7 shows that an appreciable difference exists in the delignification-digestibility response between hardwoods and softwoods. With the two hardwoods, digestibility increases rapidly with delignification and then approaches a digestibility plateau of about 90% as delignification approaches completion. With the two softwoods, there is a distinct lag phase, especially pronounced with Douglas-fir, during which extensive delignification is accompanied by only minor increases in digestibility. Following this lag phase, digestibility rises rapidly and almost linearly with delignification up to the digestibility maximum.

As interpolated from these four curves, the extent of delignification necessary to obtain a product having an in vitro digestibility of 60%, that of a good quality hay, is shown in Table VII along with data on the lignin content of the original woods and lignin content of the pulp. In common with alkali treatment (Figure 4), digestibility response strongly correlates with lignin content, response being measured in terms of the degree of pulping action needed to achieve a specified level of product utilization. Additional support for this lignin dependency was obtained by Saarinen et al. in an investigation of the in vivo digestibility of a series of birch and spruce pulps prepared by 10 different pulping techniques (32). Recalculation of

Table VII

Degree of Delignification Required to Attain

60% *in vitro* Digestibility

Wood	Required delignification[a]	Lignin in original wood	Lignin in pulp
		%	
Paper birch	25	20	21
Red oak	35	23	20
Red pine	65	27	14
Douglas-fir	73	32	13

[a]Based on original wood.

their data provided the results shown in Figure 8, which also includes curves for red pine and paper birch from Figure 7 for comparison. In spite of the wide variation in delignification techniques employed by the two investigations, the results are quite comparable. This leads to the further conclusion that it is primarily the degree of delignification that governs pulp digestibility, not the method of pulping.

A similar relationship was encountered with respect to the growth of the fungus Aspergillus fumigatus on a variety of commercial pulps prepared under different conditions (33). As determined by the protein content of the fungal mass, reasonable growth on hardwood could be obtained at lignin contents of 14% or less, whereas fungal growth on softwoods was restricted to pulps having less than 3% residual lignin.

Pulp and Papermill Residues. It is estimated that 80 lb. of fiber residues are generated for each ton of wood pulp that is produced and processed into finished products. Thus, more than 1.7 million tons per year of pulp and papermaking fiber residues are produced annually. Most of these residues have undergone at least partial delignification, which increases the accessibility of the wood carbohydrates to the rumen microorganisms and associated enzyme systems. In search for productive outlets for the fibrous residues, in vitro and in vivo estimates of digestibility and chemical analysis for lignin, total carbohydrate, and ash constituents were made on representative samples of commercial residues. On selected residues, feeding trials were conducted to observe ewe and beef steer performance (10).

Data for composition and in vitro dry matter digestibility of various types of commercially obtained pulpmill residues are given in Table VIII. As expected, groundwood fines yielded digestibility values comparable to those observed for sawdust of the same species--0% for the pine and spruce and about 35% for aspen. All of the listed screen rejects and chemical pulp fines had digestibilities of more than 40%, and digestibility of two of the pulp fines was more than 70%. Thus, based on in vitro dry matter digestibility, any of the screen rejects and chemical pulp fines could serve as a useful source of dietary energy for ruminants. The mixed hardwood, kraft bleached chemical pulp fines are essentially pure cellulose.

It can be noted in Table VIII that the Klason lignin and the total carbohydrate contents of the aspen groundwood, aspen sulfite screen rejects, and aspen sulfite parenchyma cell fines are almost identical, whereas the in vitro dry matter digestibility ranges from 37-73%. The digestibility of fines of aspen parenchyma cells, for example, is higher than would be predicted on the basis of lignin content because the parenchyma cells contain substances that analyze as lignin. Microscopic examination

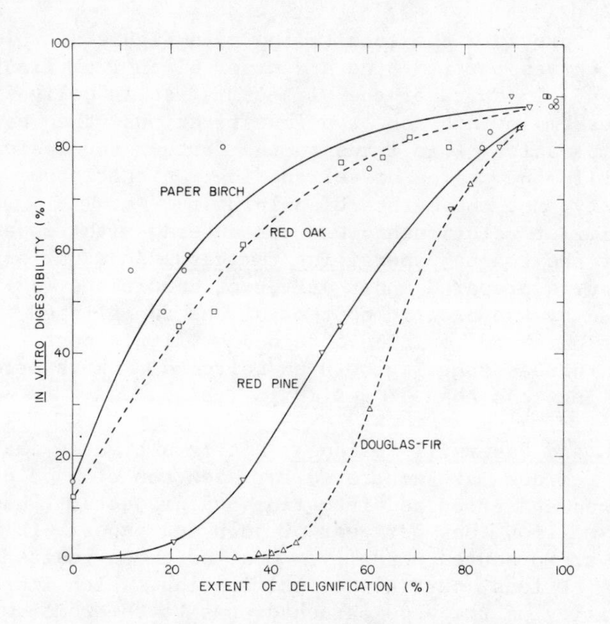

Figure 7. Relationship between in vitro *digestibility and extent of delignification for kraft pulps made from four wood species*

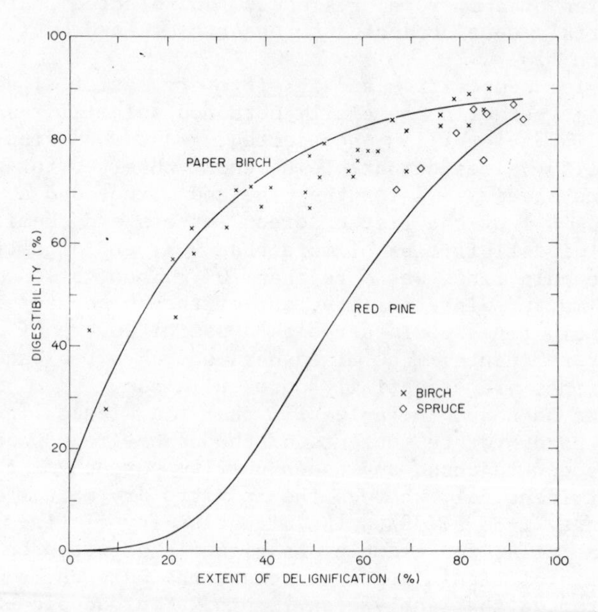

Figure 8. Relationship between digestibility and extent of delignification for wood pulps. (Data points from Saarinen, et al. *(32). Curves from Figure 7.)*

Table VIII

Composition and <u>in vitro</u> Rumen Digestibility of Pulpmill Residues

Type of residue	Lignin	Carbo-hydrate	Ash	Digesti-bility
	%			
Groundwood fines				
Aspen	21	73	1	37
Southern pine	31	59	1	0
Spruce	31	60	1	0
Screen rejects				
Aspen sulfite	19	77	2	66
Mixed hardwood, sulfite	24	65	14	54
Mixed hardwood, kraft	25	74	9	44
Chemical pulp fines				
Mixed hardwood, kraft (bleached)	<1	109	1	95
Aspen sulfite (parenchyma cells)	20	73	2	73
Southern pine, kraft (unbleached)	28	68	4	46

Table IX

Composition and *in vitro* Rumen Digestibility of Combined Pulp and Paper Mill Sludges

Type of residue	Lignin	Carbo-hydrate	Ash	Digesti-bility
	%			
Groundwood mill				
Mixed species + mixed chemical pulps	50	41	38	24
Southern pine + mixed hardwood kraft	24	60	15	19
Semichemical pulpmill				
Aspen	20	71	2	57
Aspen + mixed hardwoods	55	29	13	6
Chemical pulpmill				
Deinked waste paper, tissue	23	71	22	72
Milk carton stock	28	67	25	65
Mixed chemical pulps, tissue	17	76	13	60
Aspen and spruce sulfite	45	46	45	35

of these fines showed the presence of large quantities of dark resin-like globules. Successive extraction of these fines with ethanol and ethanol-benzene (1/2; v/v) removed more than 15% of the sample. Klason lignin content after extraction was 8.4%.

The digestibilities of the southern pine unbleached kraft pulp fines are also higher than would be predicted on the basis of lignin content. Southern pine wood and the unbleached pulp also contain substances that could analyze as lignin.

Table IX shows the composition and the in vitro dry matter digestibility of various combined pulpmill and papermill primary clarifier or lagoon sludges. Because the groundwood mill sludges are mostly groundwood fiber, the digestibility is expected to be low although the total carbohydrate content is high. The digestibility of this type of sludge will increase as the amount of chemical pulp fiber increases in the sludge. One of the semichemical pulpmill sludges was high in digestibility and total carbohydrate and low in ash, but the other was low in digestibility and total carbohydrate. This indicates the amount of variation that can be expected between mills that use the same pulping process. The digestibility of the other residues ranged from 35-72% with ash contents ranging from 13-45%.

The Klason lignin results also include acid-insoluble paper additives (ash) as lignin. Errors in the lignin analysis are evident in the data listed in Table IX for the combined pulpmill and papermill residues that have high ash content.

Composition of the ash from five pulp residues are shown in Table X, with data for aspen wood and alfalfa hay included for comparison. Except for sulfur, the residues generally exhibit lower levels of the elements P, K, Ca, Mg, and Na than does alfalfa. The Ca level in one residue is higher than that of alfalfa; the Na level is higher in two residues. Certain residues have appreciable amounts of Al and Fe. In some cases, water treatment sludges may enter the clarifiers along with the fiber residue. This would increase the levels of Al and Fe. Residue 7 is high in Zn and Mn, and residues 3, 5, and 7 are high in Cu.

A number of sludges have digestibility values comparable to hay. Their suitability for animal feed, however, will depend on the amount of ash and the chemical nature of the individual ash constituents. For example, moderate levels of clay-type filler could be tolerated, but the presence of more than trace amounts of certain heavy metals would rule out use as a feedstuff. Thus each pulp and papermaking residue should be chemically characterized before it can be recommended as a feedstuff.

Four typical residues--screen rejects from the sulfite pulping of aspen, unbleached parenchyma cell fines from an aspen sulfite tissue mill, unbleached fines from a southern pine kraft mill, and bleached fines from a mixed hardwood southern kraft mill--were blended with other ration ingredients, pelleted and fed to goats, sheep, and steers (10). Results from the

Table X

Composition of Ash From Selected Residues, Aspen Wood, and Alfalfa

Ash constituent	Type of residue[a]						
	1	2	3	4	5	6	7
	%[b]						
P	0.003	0.23	0.04	<0.01	0.02	<0.01	0.23
K	.06	2.1	.30	.05	.10	<.02	.10
Ca	.18	1.3	2.6	.70	.60	.21	.28
Mg	.03	.30	.04	.04	.08	<.01	.08
Na	.003	.15	.03	.10	.02	.43	.20
S	--	.30	1.5	1.1	.37	.28	.62
	ppm[b]						
Al	16	--	520	66	670	97	540
Ba	19	--	91	13	32	21	16
Fe	35	200	2,300	63	340	95	350
Sr	10	--	44	13	11	6	16
B	3	--	16	2	4	4	14
Cu	6	13	74	6	40	8	99

(Page 1 of 2)

Table X

Composition of Ash From Selected Residues, Aspen Wood, and Alfalfa--continued

Ash constituent	Type of residue[a]						
	1	2	3	4	5	6	7
	ppm[b]						
Zn	19	20	14	6	37	4	330
Mn	10	32	44	2	7	9	330
Cr	.3	--	7	1	5	5	13
	%						
Total ash[b]	0.60	8.0	17.4	2.1	2.1	1.8	3.4

[a]1, aspen wood; 2, alfalfa hay; 3, mixed hardwood sulfite screen rejects; 4, aspen sulfite screen rejects; 5, aspen sulfite parenchyma cell fines; 6, mixed hardwood sulfite pulp fines; and 7, southern pine unbleached kraft pulp fines.

[b]Based on moisture-free sample.

(Page 2 of 2)

digestibility trials indicate that the *in vivo* dry matter digestibilities are 58, 52, 47, and 78%, respectively. This indicates substantial utilization of the carbohydrate constituents.

The rumen contents of steers fed unbleached southern pine kraft mill fines and steers fed a control ration containing no pulp fines were analyzed for pH, ammonia, volatile fatty acids, and microbial population. No significant differences could be observed between the rumen contents of steers on the control ration and those on the experimental rations.

Steers averaging 226 kg., fed a ration containing 50% unbleached southern pine kraft mill fines, gained 0.5 kg. per day during a 58-day growth trial. During another growth trial, steers averaging 221 kg. were fed a ration containing 75% parenchyma cell fines. These steers gained an average of 0.45 kg. per day during 101 days. These weight gains are not high but they are acceptable wintering growth rates. Feed efficiencies for the two experiments were 11.7 and 16.9 kg. feed per kg. gain.

Rations containing 60% and 75% parenchyma cell fines have been fed to ewes and beef cows with good results. Ewes fed pelleted rations containing 75% fines for one year, and supplemented with additional grain during the last month of pregnancy and during lactation, produced as much wool and weaned as many lambs as did a hay fed control group. Ewes fed a similar ration containing aspen bark in place of pulp fines performed equally as well. Total feed consumption was higher for the groups fed pulp fines and aspen bark reflecting a slightly lower digestibility of these materials compared to hay.

Beef cows fed 2-3 kg. of hay plus a mixture of parenchyma cell fines and grain (83% fines and 17% of grain and mineral supplement) for a period of about 7 months appeared normal in every respect. Palatability of the pulp fines mixture was good.

Summary

The roughage qualities of wood in ruminant rations have been evaluated and compared to other roughages. Wood has been shown to be effective as a roughage replacement. Depending upon the other ration ingredients, concentrations of 5-15% screened sawdust in rations for beef cattle appears practical. For lactating dairy cows, aspen sawdust could be used as a roughage extender or as a partial roughage substitute in high grain rations. Some long hay appears to be necessary in the ration to stabilize feed intake.

The potential of wood and bark, chemically and physically treated wood, and pulp and papermaking residues as energy sources in ruminant rations has been examined by chemical analysis and *in vitro* and *in vivo* methods. *In vitro* rumen and enzyme methods were developed to assay wood-based materials for digestibility.

Of the woods tested, all of the coniferous species are essentially undigested by rumen micro-organisms. Deciduous species, with a few exceptions, are only slightly digested. Aspen is the most highly digestible species tested, giving both an *in vitro* and *in vivo* digestibility of about 35%. Aspen bark is about 50% digestible. The resistance to micro-organisms appears to be related to the lignin-carbohydrate complex and the crystallinity of the cellulose.

The coniferous species and most deciduous species were quite resistant to vibratory ball milling, electron irradiation, dilute alkali, and liquid ammonia treatments to increase digestibility. Treatment with gaseous sulfur dioxide appears especially interesting as a way to increase the digestibility of wood. Since no water is added and the product is not washed, yields of over 100% are obtained. The product was accepted by animals during digestion trials.

Delignification of wood by normal wood pulping methods produces materials with high rumen digestibility. It was shown that the digestibility of deciduous species increases rapidly compared to coniferous species as lignin is removed. It was also shown that digestibility depends upon the extent of lignin removal and not upon the method of lignin removal.

Pulp and papermaking residues were analyzed for lignin, carbohydrate, rumen digestibility, ash, and ash constituents. *In vitro* rumen digestibility of many of the residues ranged from 45-60%; some attained levels as high as 90%. *In vivo* digestibilities of four typical pulpmill residues ranged between 47 and 78%, and were in reasonable agreement with the *in vitro* values. Certain residues appear suitable as feed ingredients while others are not suitable because they contain high amounts of ash or contain ash with high concentrations of heavy metal contaminants.

Pulp fines, constituting 50-75% of the ration for steers, ewes, and beef cows were readily consumed. Steer growth rates of approximately 0.5 kg. per day were obtained. Ewes and cows were maintained at an adequate level of nutrition so normal reproduction occurred and growth of nursing offspring was normal. Total feed consumption tended to be higher with the groups fed wood residues, reflecting the slightly lower digestibility of these materials compared to hay.

Literature Cited

1. Sherrard, E. C., and Blanco, G. W. J. of Ind. and Eng. Chem. (1921) 13 61-65.
2. Morrison, F. B., Humphrey, G. C., and Hulce, R. S. Unpublished report. Forest Products Lab. (1922).
3. Woodward, F. E., Converse, H. T., Hale, W. R., and McNulty, J. B. U.S.D.A., Dept. Bull. No. 1272 (1924) 9-12.
4. Archibald, J. G. J. of Dairy Sci. (1926) 9 257-271.
5. U.S.D.A., F.S., Forest Products Lab., Report No. 1731 (1955).
6. Lloyd, R. A., and Harris, J. F. U.S.D.A., F.S., Forest Products Lab. Report No. 2029 (1955).
7. Harris, E. E., Hajny, G. J., and Johnson, M. C. Ind. and Eng. Chem. (1951) 43 1593-1596.
8. Turner, H. D., Forest Prod. J. (1964) 14 (7) 282-284.
9. Scott, R. W., Millett, M. A., and Hajny, G. J. Forest Prod. J. (1969) 19 (4) 14-18.
10. Millett, M. A., Baker, A. J., Satter, L. D., McGovern, J. N., and Dinnius, D. A. J. Animal Sci. (1973) 37 599-607.
11. Mellenberger, R. W., Satter, L. D., Millett, M. A., and Baker, A. J. J. Animal Sci. (1970) 30 1005-1011.
12. Moore, W. E., Effland, M. J., and Millett, M. A. J. Agr. Food Chem. (1972) 20 1173-1175.
13. Millett, M. A., Baker, A. J., Feist, W. C., Mellenberger, R. W., and Satter, L. D. J. Animal Sci. (1970) 31 781-788.
14. Dinius, D. A., and Baumgardt, B. R. J. Dairy Sci. (1970) 53 311-316.
15. Mellenberger, R. W., Satter, L. D., Millett, M. A., and Baker, A. J. J. Animal Sci. (1971) 32 756-763.
16. Baumgardt, B. R. Feedlot (April 1969).
17. Anthony, W. B., Cunningham, J. P., and Harris, R. R. "Cellulases and Their Applications" pp. 470. American Chemical Soc., Washington (1969).
18. Dinius, D. A., Peterson, D., Long, T. A., and Baumgardt, B. R. J. Animal Sci. (1970) 30 309-312.
19. El-Sabbon, F. F., Long, T. A., and Baumgardt, B. R. J. Animal Sci. (1971) 32 749-755.
20. Van Soest, P. J. J. Dairy Sci. (1963) 46 204-216.
21. Satter, L. D., Baker, A. J., and Millett, M. A. J. Dairy Sci. (1970) 53 1455-1460.
22. Satter, L. D., Lang, R. L., Baker, A. J., and Millett, M. A. J. Dairy Sci. (1973) 56 1291-1297.
23. Mellenberger, R. W., Satter, L. D., Millett, M. A., and Baker, A. J. J. Animal Sci. (1971) 32 756-763.
24. Millett, M. A., Baker, A. J., and Satter, L. D. Paper presented at Special Seminar on Cellulose as a Chemical and Energy Resource, Univ. of Calif., Berkeley, June 25-27, 1974. (Proceedings to be published).

25. Kirk, T. K., and Moore, W. E. Wood and Fiber (1972) 4 (2) 72-79.
26. Saeman, J. F., Millett, M. A., and Lawton, E. J. Ind. Eng. Chem. (1952) 44 2848-2851.
27. Segel, L., Loeb, L., and Creely, J. J. J. Polymer Sci. (1954) 13 193-206.
28. Tarkow, H., and Feist, W. C. "Cellulases and Their Applications" pp. 470. American Chemical Soc., Washington (1969).
29. Tarkow, H., and Feist, W. C. Tappi (1968) 51 (2) 80-83.
30. Feist, W. C., Baker, A. J., and Tarkow, H. J. Animal Sci. (1970) 30 832-835.
31. Baker, A. J. J. Animal Sci. (1973) 36 768-771.
32. Saarinen, P., Jensen, W. J., and Alhojärvi, J. Acta. Agral. Fennica (1959) 94 41-62.
33. Baker, A. J., Mohaupt, A. A., and Spino, D. F. J. Animal Sci. (1973) 37 179-182.

7

Enzymes: Specific Tools Coming of Age

DONALD F. DURSO and ANITA VILLARREAL

Department of Forest Science, College of Agriculture, Texas A&M University, College Station, Tex. 77840

Abstract

Cellulose chemistry has long represented a difficult battleground for researchers who have successfully applied science to other fields. The seeming simplicity and the true complexity of the cellulose "package"---the fiber structure---have served to entrap those who would unravel its mysteries and use if efficiently. Combining the mysteries of cellulose and enzymes has produced an area containing little fact and much fiction.

The studies of this paper provide methods to predict enzymatic hydrolysis of cellulose and indicate a practical method for rapid conversion of cellulosic substrates into glucose.

* * *

As part of a project aiming at some useful disposition for mesquite, means have been devised for the production and the characterization of a potent cellulase system obtained as an exo-enzyme from Trichoderma viride(1,2). A sample of the enzyme and the preferred strain of T. viride were kindly supplied by Mary Mandels several years ago. The methods to produce the cell-free enzyme and to measure its activity are essentially those revealed by Mandels (1,3).

The results presented in this paper stem from an effort to standardize the enzyme system so that it can be used as a reliable, reproducible yardstick to measure progress in the project to make mesquite wood totally accessible to cellulases. Although this work is based on pure cellulose (filter paper) and cell-free enzyme, the conclusions appear to be applicable to the digestion of lignocellulose material by intact organisms in the rumen of cattle and sheep (4).

Methods devised for the concentration of the enzyme, the measurement of its activity via DNS color reaction and the assay of its action on filter paper are outlined in Appendix B. With the standard enzyme concentration, the rate of weight loss is shown as the "Normal" curve in Fig. 1. The non-linear effect of changing only the enzyme concentration is also indicated. Here the response seems to be logarithmic, agreeing

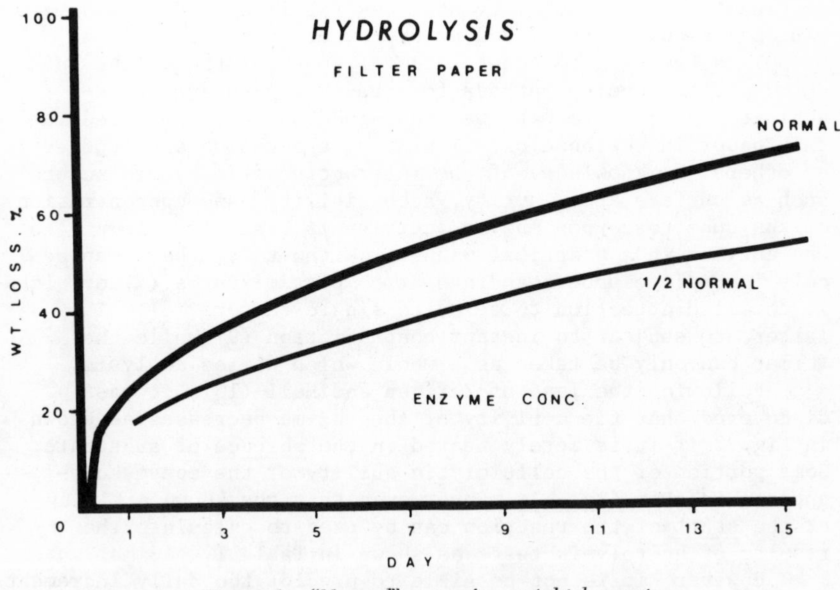

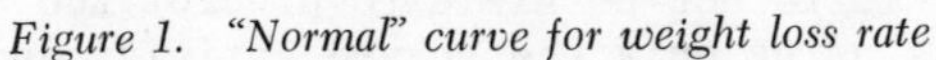
Figure 1. "Normal" curve for weight loss rate

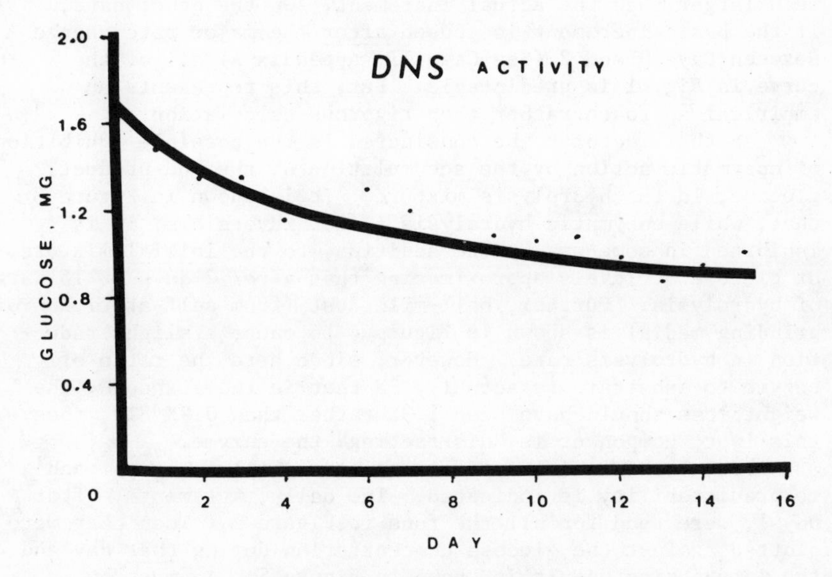

Figure 2. Decrease in enzyme activity

with some (5) while disagreeing with others (6); in any case, setting the enzyme activity at a desired level (DNS = 2.0 mg of glucose) requires trial and error rather than calculation.

In attempting to understand the shape of the weight loss curve and to develop methods for predicting substrate reactivity, certain results were obtained and are presented in this paper in the hope of stimulating discussion and studies by others. A knowledge of the interactive effects of factors such as surface area, purity, accessibility, and concentration of the substrate upon enzyme activity is basic in attempts to use enzymes in a practical manner. Without it, there can be only incomplete understandings of complex mixtures (7) or highly detailed attention to selected single factors (8). The latter are subject to instant contradiction (9) while the former can only be taken as a whole which defies analysis.

Following the lead of Zeffren and Hall (10), it was discovered that the activity of the enzyme decreases as shown in Fig. 2 if it is merely heated in the absence of substrate. Some portion of the cellulolytic ability of the enzyme disappears in a predictable manner such that the known activity at the start of the reaction can be used to calculate the results of hydrolysis tests as shown in Table 1.

However, it is not possible to predict the daily increment of weight loss using only the enzyme activity and the substrate concentration at any given day. In addition, if one calculates the incremental increase in weight loss by an apparently logical method (See Case I, Appendix A), it is found to be much larger than the actual increment. On the other hand, if the basic increment is chosen after the major rate change between Days 0 and 2 (See Case II, Appendix A) all of the curve in Fig. 1 is predictable. But, this represents an empirical approach rather than rigorous calculation.

Another factor to be considered is the possible inhibition of enzymatic action by the accumulation of the end-product, glucose, in the hydrolysis mixture. It is shown in Figure 3 that, while enzymatic hydrolysis is not reversible, it is inhibited in some way by the addition, to the initial mixture, of glucose at levels approximating that after 2 days or 15 days of hydrolysis. Further, ball-mill dust (from self-attrition of grinding media) is shown in Figure 4 to cause a slight reduction in hydrolysis rate. However, since here the ratio of enzyme to substrate is actually 2X that in the standard, the weight loss should have been 1.3X rather than 0.9X STD; thus, this inert component is "distracting" the enzyme.

From these inhibition data, another empirical approach to predictability is indicated. The daily increments, after Day 1, were read for all the runs in Figure 3. Then they were plotted against the glucose concentration during that day and the interesting result is shown in Figure 5. It must be remembered that different substrate and glucose concentrations,

Table I. Enzyme Activity on Filter Paper

REACTION TIME, DAYS	ENZYME CONDITION** AT START OF HYDROLYSIS VS. WT. LOSS						
	FRESH, STD.	4-DAY**		10-DAY**		15-DAY**	
		C*	F	C*	F	C*	F
4	41.1	29.5	29.4	-	-	21.8	22.5
7	53.1	-	-	-	-	28.1	28.0
10	59.9	-	-	-	-	31.7	30.9
15	70.2	50.4	47.4	41.3	38.0	37.2	35.4
DNS ACTIVITY	1.7	1.22		1.0		0.9	

* CALCULATED = STD. VALUE X DNS ACTIVITY ÷ 1.7

** PRE-HEAT DAYS AT 50°C, pH 5 IN SEALED CONTAINER

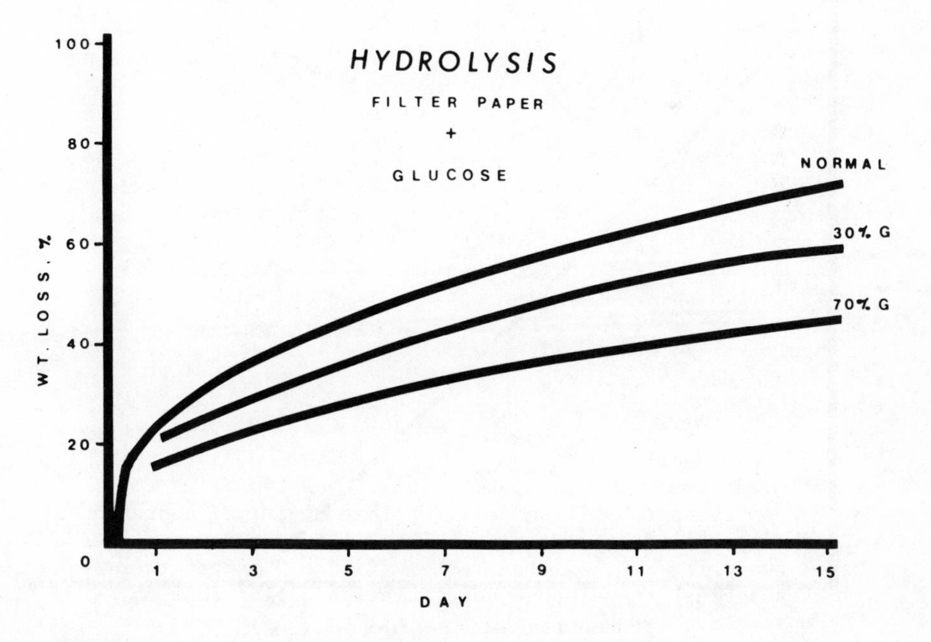

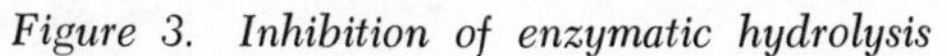

Figure 3. Inhibition of enzymatic hydrolysis

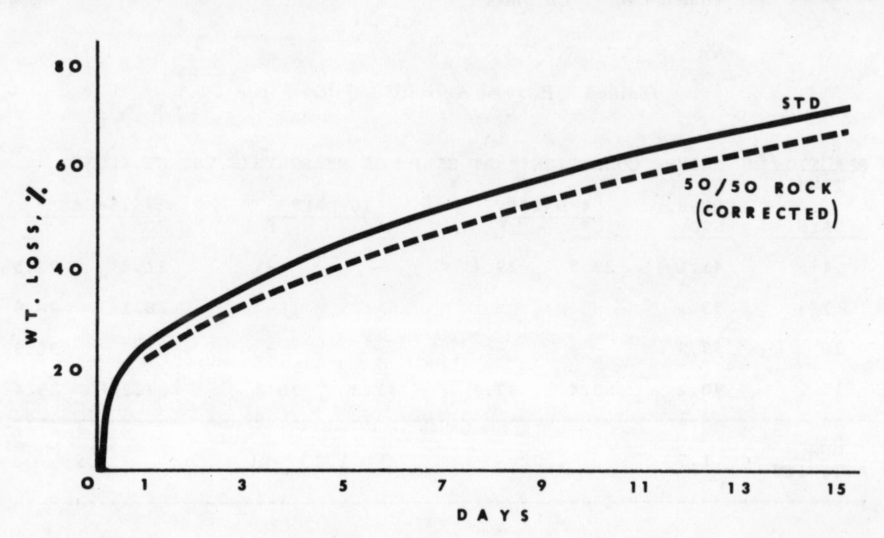

Figure 4. Cellulase hydrolysis of filter paper.Effect of ball-mill "rock"

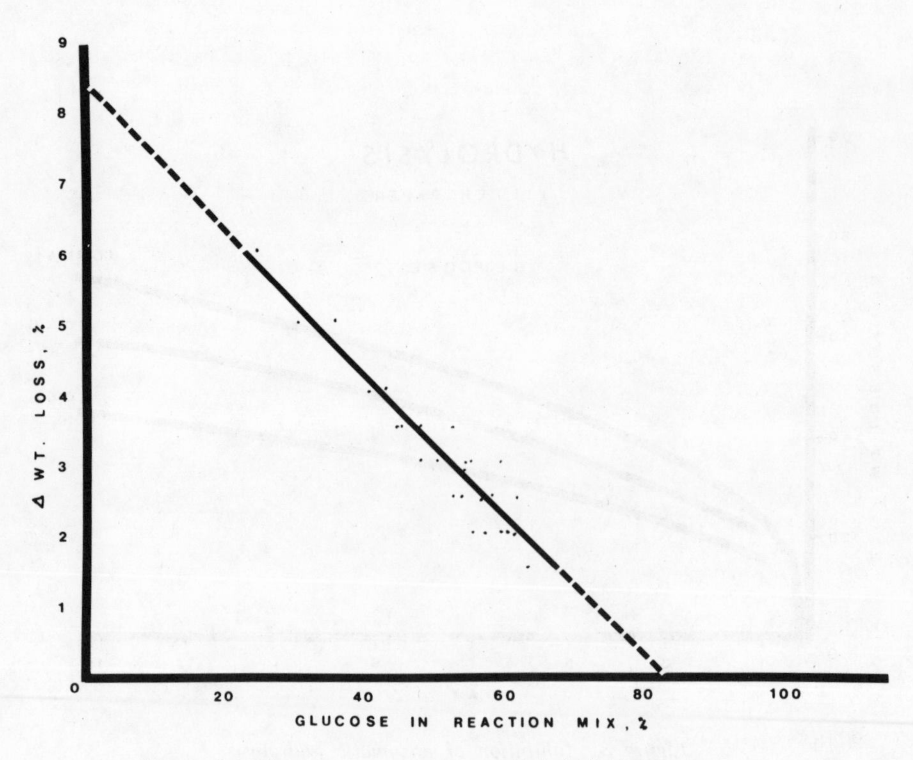

Figure 5. Glucose concentration vs. daily hydrolysis increment

and enzyme activities are included. Yet, using only the glucose concentration at any given degree of hydrolysis to read the next daily weight loss increment, it becomes possible to predict the entire "Normal" curve of Figure 1 from Day 1 through Day 15. Further, it is interesting to note that this relation (Figure 5) predicts the cessation of hydrolysis at the 80% level.

This Figure 5 also led to the conclusion that something vitally decisive happens during the first day of hydrolysis. Therefore, the reactions were carried out at shorter intervals and the data are given in Figure 6.

It is clear that a remarkable decrease in substrate degradation occurs after the first hour. In this first hour, there has occurred only a minor change in each of the primary factors (DNS enzyme activity and substrate concentration) and the glucose effect is minor (note that even 70% added glucose only slightly affects the initial hydrolysis rate); yet the hourly increment drops rapidly from almost 9% to much less than 1%.

In the hopes of better defining the possible effects of substrate structure or enzyme destruction (absorption?), the experiments summarized in Figure 7 were carried out. The standard was continued for 5 hours while each of 12 other reactions were given alternate treatments at the end of the 3rd hour. In A, the residue was recovered by filtration and washing; then, assuming about 10% weight loss, it was redispersed in 90% of the initial volume of enzyme to restore the original substrate to enzyme ratio. This was done in duplicate and the reaction progress measured at the end of 1 and 2 more hours. In B, 100% fresh enzyme was added to selected tubes and again these were analyzed at 1 and 2 hours. In C, 100% fresh substrate was added and then the analyses were carried out as before.

In Figure 7, A indicates that enzymatic hydrolysis had been stymied by some attack on a factor in the enzyme rather than in the substrate. B and C are somewhat confused by the presence of glucose and "old" substrate and enzyme; however, from the appearance of the reaction mix it can be deduced that reaction had ceased in C after about 1 hour while B still continued at a renewed pace. It is thus inferred but not proven that some secondary effect of substrate on the enzyme system is mainly responsible for the loss of the initial rate of reaction. Substrate structure per se, if important is minor under the conditions of these experiments.

Conclusions

There are some positive feelings from these studies despite the doubts and questions which require further work. It is possible by rigorous and empirical methods, combined, to

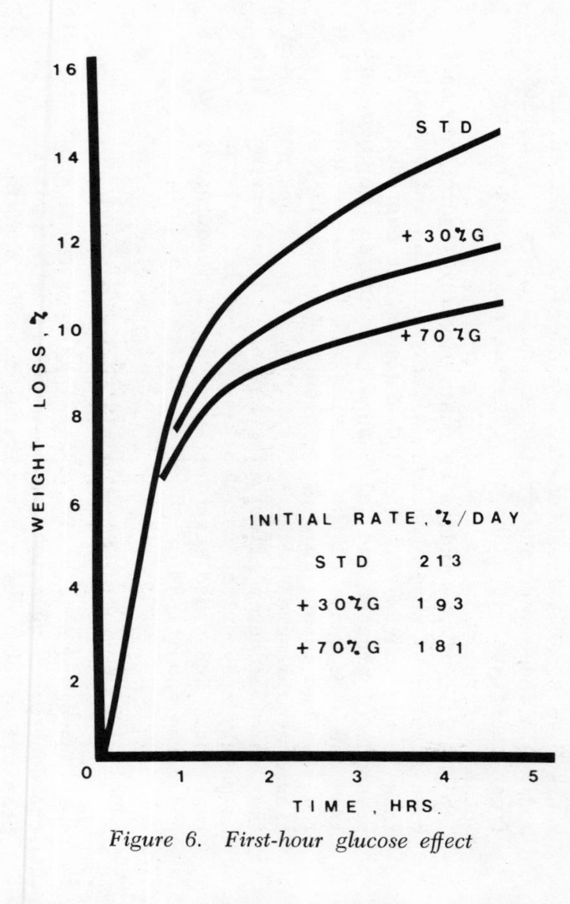

Figure 6. First-hour glucose effect

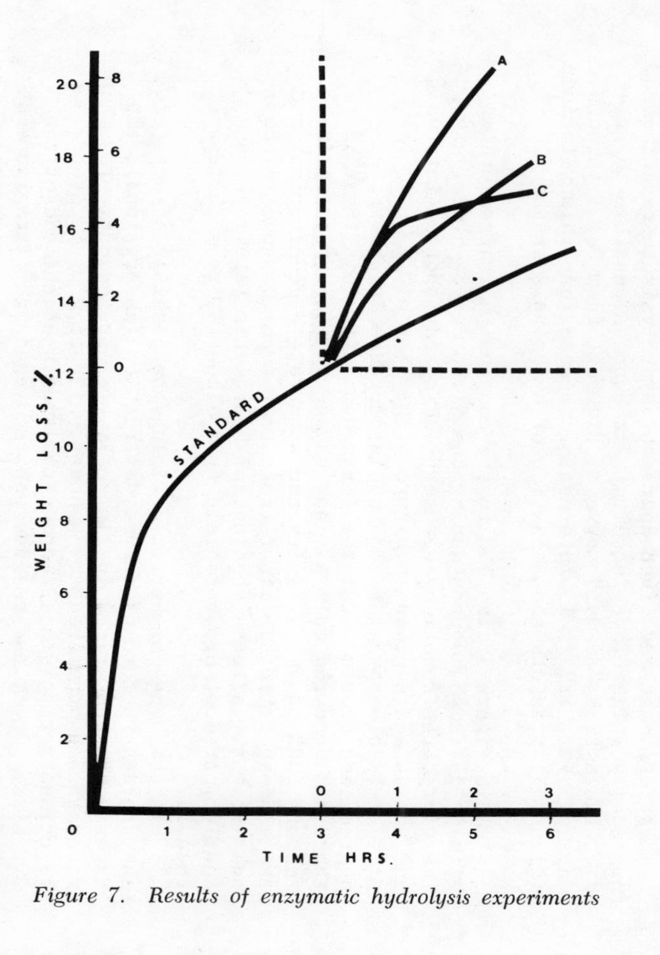

Figure 7. Results of enzymatic hydrolysis experiments

predict the rate of hydrolysis of cellulose from about 10% reaction to its end. It is shown that glucose, rock and other attractions for the enzyme will distract it from its purpose.

It appears these findings can be put to use for the production of glucose at very practical rates, if one can maintain the initial conditions. These can be obtained by

1. Dialysis to maintain zero glucose in the reactor.
2. Addition of substrate at the rate it is hydrolyzing.
3. Addition of fresh enzyme to correct for activity loss.

On this basis the data in this paper can be used by practical chemists or chemical engineers to design a reactor into which one would add continuously the raw materials and continuously remove the glucose through a suitable membrane.

If there is need for such glucose, economic and pilot plant studies can now begin to develop this system. It will give continuous reduction of carbohydrate wastes to glucose in hours, rather than the usually prescribed days of reaction time (11).

Literature Cited

1. Villarreal, Anita, The Action of T. viride Cellulase on Purified and Partially Purified Cellulosic Substrates, M.S. Thesis, Texas A&M University, December 1972.
2. Goldstein, Irving S. and Villarreal, Anita, Wood Sci. (1972), 5, 15-20.
3. Mandels, Mary and Weber, James, The Production of Cellulases, in "Cellulases and their Applications, Advances in Chemistry Series 95", 391-414, American Chemical Society, Washington, D.C., 1969.
4. Mellenberger, R.W., et al, J. Animal Sci. (1970), 30, 1005-1011.
5. Reese, E.T. and Mandels, Mary, Enzymatic Degradation, in "Cellulose and Cellulose Derivatives, Part V", 1088, Wiley-Interscience, New York, 1971.
6. Halliwell, G., Measurement of Cellulase and Factors Affecting its Activity, in "Enzymic Hydrolysis of Cellulose and Related Materials," 83, Pergamon, London, 1963.
7. Davis, C.L., et al, Conference on Cellulose Utilization in the Rumen, Federation Proc. (1973), 32, 1803-1825.
8. Caulfield, D.F. and Moore, W.E., Wood Sci. (1974), 6, 375-379.
9. McKeown, J.U. and Lyness, W.I., J. Polymer Sci. (1960), 47, 9-17.
10. Zeffren, E. and Hall, P.L., Kinetics I & III, in "The Study of Enzyme Mechanisms," 62; 63-67; 87-95, Wiley-Interscience, New York, 1973.
11. Anon., C&EN, May 27, 1974, page 20; Moore, W.E., et al, J. Agr. Food Chem. (1972), 20 (6), 1173-5.

Appendix A

Calculation of Daily Weight Loss Increment (Example: Day 7 to Day 8).

I. Using Day 0 to Day 1 as the basic increment
Enzyme Activity (DNS): Initial = 1.7; Day 7 = 1.08
Substrate Concentration: At Day 7 = 48% of Original
Initial Increment = 25% Weight Loss

Calculated Day 7 to Day 8 Increment
= Initial Increment X Relative Substrate Conc. X Enzyme Activity
= 25 X 0.48 X 1.08/1.7
= 7.62%

II. Using Day 2 to Day 3 as the basic increment
Enzyme Activity (DNS): Day 2 = 1.38; Day 7 = 1.08
Substrate Conc.: At Day 2 = 69%; At Day 7 = 48%
Basic Increment = 37-31 = 6% Daily Weight Loss

Calculated Day 7 to Day 8 Increment
= 6 X 48/69 X 1.08/1.38
= 3.27%

Note: Using the case II procedure, the entire Normal weight loss curve of Figure 1 can be constructed from Day 4 through Day 15.

Appendix B

Enzyme Preparation. Per the procedure in Reference 3 except that 14.7 g/l of sodium citrate were added to the nutrient medium.

Enzyme Concentration. If the enzyme solution needs to be concentrated, the Amicon ultrafiltration apparatus is used. Nitrogen is used as the source of pressure. The filter used is an Amicon membrane (PM 30, 62mm). The enzyme solution is transferred to the reservoir tank. The nitrogen is connected to the inlet of the reservoir tank. The outlet connection is attached to the filtering cell which has the membrane in place (glossy side up). The outlet hose of the filtering cell, is allowed to sit in a collecting flask.

The reservoir tank is pressured to 60 psi with nitrogen and then the nitrogen source is removed. The stirrer is turned on at a speed which produces a vortex of 1/3 the volume in the cell. When the desired volume has been collected, the stirrer is stopped and the vent of the reservoir tank is released slowly. After the pressure is completely down, the stirrer is turned on at a slow speed for about 5 minutes. The concentrated enzyme in the cell and the tank is saved and refrigerated. Adjust the pH to 4.8-5.0 before using.

Cellulase Assay by DNS Color Reaction

Per the procedure in Reference 3.

Cellulase Assay by Weight Loss Method

Apparatus

Screw-cap culture tubes, 30 ml. 20x150 mm
Tared sintered glass crucibles, 30ml, coarse porosity
50^oC H_2O bath
Vortex mixer

Reagents

Trichoderma viride cellulase adjusted to a pH of 5.0 and approximate activity of 2.0 mg glucose (DNS method). The DNS activity level is obtained by concentration or dilution.

Procedure

1. Weight out 0.5 g of 1x6 cm Whatman #1 filter paper strips. Place in screw-cap culture tubes (sterilization not necessary) as intact strips.
2. Add 10 ml of *T. viride* cellulase (pH 5.0 and 2.0 mg activity).
3. Incubate at 50^oC for 15 days and mix daily on a vortex mixer.
4. At the end of the incubation period, filter the residue on a tared sintered glass crucible, rinse well with cold water, and oven dry at 105^oC. Determine the weight loss as a percentage of the original sample weight.

8

Reaction of Alkylene Oxides with Wood

ROGER M. ROWELL

Forest Products Laboratory, P.O. Box 5130, Madison, Wis. 53705

All of the commercial wood preservatives presently used in the United States are effective in preventing attack by microorganisms because of their toxic nature. Because of the concern these chemicals have on the environment, alternative methods based on nontoxic procedures are being investigated.

Chemical modification as a possible preservative treatment for wood is based on the theory that enzymes (cellulase) must directly contact the substrate (cellulose) and this substrate must have a specific configuration. If the cellulosic substrate is chemically changed, this highly selective reaction cannot take place. Chemical modification can also change the hydrophilic nature of wood. In some cases water, a necessity for decay organisms, is excluded from biological sites. The chemicals used for modification need not be toxic to the organism because their action renders the substrate unrecognizable as a food source to support microbial growth. For wood preservation, this means that it is possible to treat wood in such a manner that attack by wood-destroying fungi will be prevented and the material will be safe for humans to handle. For wood usages in which human contact is essential, nontoxic preservatives may well be specified or required in the future. An added benefit of most chemical modification treatments to wood is the resulting bulking action gives the treated wood very good dimensional stability. The objective of this research area is to develop a permanent, nonhazardous preservative based on the chemical reactivity of the wood components.

Several requirements must be met for a successful treating system.

Of the thousands of chemicals available, either commercially or by synthetic means, most can be eliminated because they fail to meet the properties listed below:

1. The chemical must contain functional groups which will react with hydroxyl groups of the wood components. This may seem obvious to most, but there are many literature citings of chemicals that fail to react with wood components when, in fact, they

did not contain functional groups which could react. They should never have been tried in the first place.

2. The overall toxicity of the chemicals must be carefully considered. The chemicals need have no toxicity to the microorganisms, should not be toxic or carcinogenic to humans in the finished product, and should be as nontoxic as possible in the treating stage. The chemical should be as noncorrosive as possible to eliminate the building of special stainless steel or glass-lined treating equipment.

3. In considering the ease with which excess reagents can be removed after treatment, a liquid with a low boiling point is advantageous. If a gas system is used, a low level of chemical substitution is usually achieved and there are problems in pressurized gas handling. Likewise if the boiling point is too high, it will be very difficult to remove after treatment. It is generally true that the lowest member of a homologous series is the most reactive and will have the lowest boiling point. The boiling point range to be considered is 30°-150° C.

4. In whole wood, accessibility of the treating reagent to the reactive chemical sites is a major consideration. To increase penetration and accessibility, the chemical system must swell the wood structure. If the reagents do not swell the wood, then another chemical or co-solvent can be added to meet this requirement.

5. Almost all chemical reactions require a catalyst. With wood as the reacting substrate, strong acid catalyses cannot be used as they cause extensive degradation. The most favorable catalyst from the standpoint of wood degradation is a weakly alkaline one. The alkaline medium is also favored as in many cases these chemicals swell the wood structure and give better penetration. The properties of the catalyst parallel those of reagents, i.e., low boiling point liquid, nontoxic, effective at low temperatures, etc. In most cases, the organic tertiary amines are best suited.

6. The experimental reaction conditions which must be met in order for a given reaction to go is another important consideration. The temperature required for complete reaction must be low enough so there is little or no wood degradation, i.e., less than 120° C. The reaction must also have a relatively fast rate of reaction with the wood components. It is important to get as fast a reaction as possible at the lowest temperature without wood degradation.

The moisture present in the wood is another consideration in the reaction conditions. It is impracticable to dry wood to less than 1% moisture, but it must be remembered that the OH group in water is more reactive than the OH group available in wood components, i.e., hydrolysis is faster than substitution. The most favorable condition is a reaction which requires a trace of moisture and the rate of hydrolysis is relatively slow.

Another consideration in this area is to keep the reaction simple. Avoid the multicomponent systems that will require complex separation after reaction for chemical recovery. The optimum here would be if the reacting chemical swells the wood structure and is the solvent as well.

7. There should be no byproducts produced during the reaction that have to be removed. If there is not a 100% reagent skeleton add-on, then the chemical cost is higher and may require recovery of the byproduct for economical reasons.

8. The chemical bond formed between the reagent and the wood components is of major importance. For permanence, this bond should have great stability to withstand weathering. In order of stability, the types of covalent chemical bonds that may be formed are: ethers > acetals > esters. The ether bond is stable to acids and bases; the acetals to bases but labile to acids and esters are labile to both acids and bases. It is obvious that the ether bond is the most desirable covalent carbon-oxygen bond that can be formed. These bonds are more stable than the glycosidic bonds between sugar units in the wood polysaccharides so the wood polymers would degrade before the grafted ether.

All of these bond possibilities consider only covalent bonding with hydroxyl groups; however, other types of chemical attachments are possible. For example, hydrogen bonding, ionic interactions, complexing, chelation, and encapsulation are all possibilities but less permanent.

9. The hydrophobic nature of the reagent needs to be considered. The chemical added to the wood must not increase the hydrophilic nature of the wood components. If the hydrophilicity is increased, the susceptibility to micro-organism attack increases. The more hydrophobic the component can be made, the better the substituted wood will withstand dimensional changes in the presence of moisture.

10. Single site substitution versus polymer formation is another consideration. The greater the degree of chemical substitution (D.S.) of wood components, the better it is for rot resistance. So, for the most part, a single reagent molecule that reacts with a single hydroxyl group is the most desirable.

Crosslinking can occur when the reagent contains reactive groups which substitute two hydroxyl groups. Crosslinking can cause the wood to become more brittle so reagents in this class must be chosen carefully.

Polymer formation within the cell wall attached to wood components gives good biological resistance and the bulking action of the polymer gives the added property of dimensional stabilization. The disadvantage of polymer formation is that a higher level of chemical add-on is required for the biological resistance than in the single site reactions.

11. The treated wood must still possess the desirable properties of wood. The strength must remain high, no or little

change in color (unless this is a desired requirement), good electrical insulation, not dangerous to handle, gluable, and finishable. In this study, the goal of chemical modification is to increase the decay resistance and the dimensional stability of wood. Chemical modification can also be used to give additional improvements such as resistance to corrosion, ultraviolet degradation, and fire.

12. A final consideration is, of course, the cost. In the laboratory experimental stage, it is not a major factor due to the high cost of chemicals when produced on a small scale. For commercialization of a chemical modification for wood, the chemical cost is a very important factor. On today's market, the limit of chemical cost of treated wood for rot resistance cannot exceed 50¢ per cubic foot. In specialty markets where dimensional stabilization is also a requirement, the chemical cost can be 2-3 times higher.

In summary, the chemicals to be laboratory tested must be capable of reacting with wood hydroxyls under neutral or mildly alkaline conditions at temperatures below 120° C. The chemical system should be simple and capable of swelling the wood structure to facilitate penetration. The complete molecule must react quickly to the wood components yielding stable chemical bonds and the treated wood must still possess the desirable properties of the wood.

One reaction system which meets the requirements is the base catalyzed reaction of alkylene oxides with hydroxyl group.

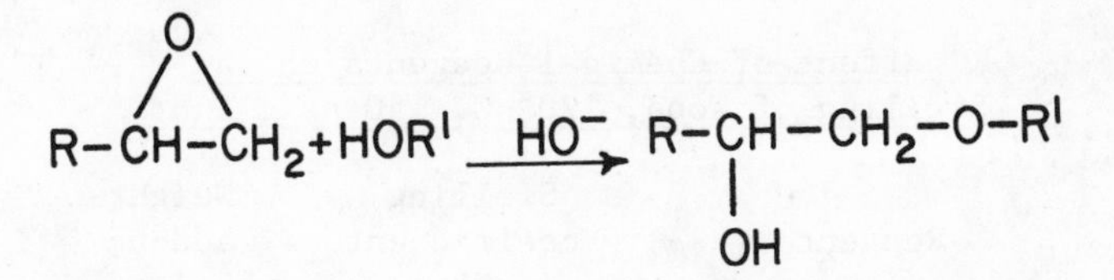

The reaction is fast, complete, generates no byproducts, forms stable ether bonds, and is catalyzed by a volatile organic amine. After the initial reaction, a new hydroxyl group originating from the epoxide is formed. From this new hydroxyl, a polymer begins to form. Due to the ionic nature of the reaction and the availability of alkoxyl ions in the wood components, the chain length is probably short due to chain transfer.

Considering the alkylene oxides or epoxides in light of the preceeding requirements, the lowest member in the series (ethylene oxide) is a gas at room temperature. Ethylene oxide

Reagent	Boiling point °C.
Ethylene oxide	10.7
Propylene oxide	35
Butylene oxide	63
Epifluorohydrin	85
Epichlorohydrin	116
Epibromohydrin	135
Trimethylamine	2.9
Triethylamine	90

catalyzed with trimethylamine have been used to react with cellulose, but high pressure equipment must be used. Propylene oxide or butylene oxide are liquids at room temperature and both can be catalyzed with triethylamine. Of the substituted halogen epoxides, epifluorohydrin is preferred but its cost prohibits its use ($8/g). The boiling point of epichlorohydrin is higher but can be easily removed from the wood after reaction.

To determine if the reaction system was capable of swelling the wood structure, the swelling coefficient of each separate reagent was determined. A southern pine block (3/4" x 3/4" x 4") was immersed in each separate reagent for 1 hour, 150 psi at 120° C. The block volume was determined ovendry before treatment and wet immediately after treatment. The weight gain during

Effect of Chemical Reagents on the Swelling of Wood, 120° C, 150 psi, 1 Hr.

Reagent	Swelling coefficient S	Weight add-on %
Water	10	0
Triethylamine	.7	.2
Propylene oxide	5.6	3.8
Epichlorohydrin	5.8	7.6

treatment is the difference between ovendry weight before treatment and ovendry again after treatment. It might be expected that the alkaline amine would swell the wood as does amines such as pyridine; however, triethylamine does not swell wood. The swelling ability of propylene oxide and epichlorohydrin are about 60% that of water. So in the epoxide reaction system, it is the epoxide that swells the wood structure.

The amount of catalyst needed was determined by reacting southern yellow pine with varying ratios of epoxide/amine. From this data,

Ratio PO/TEA	Wt. % Add-On
20/80	20
50/50	44
80/20	52
90/10	53
95/5	50
97/3	40

a ratio of 95/5 epoxide to catalyst was chosen for maximum reaction with minimum reagent recovery. The conditions of 120° C. at 150 psi were chosen for all runs. By varying the reaction time, samples were prepared with polymer add-on levels of from 7 to 60% by weight.

Changes in Volume of Southern Yellow Pine After Treatment with Alkylene Oxides

Compound	Green volume $In.^3$	Ovendry volume before $In.^3$	Ovendry volume after $In.^3$	Weight add-on %
Propylene oxide	3.48	3.24	3.42	15.9
Butylene oxide	3.60	3.24	3.60	21.1
Propylene oxide	3.66	3.42	3.66	26.1
Propylene oxide	3.60	3.30	3.66	34.1
Epichlorohydrin	3.60	3.36	3.72	41.0

At a weight percent add-on of approximately 20%, the volume of the treated wood is equal to the untreated green volume. Above about 30% weight add-on, the volume of the treated wood is larger than that of the green wood. These results show that the polymer is located in the cell wall. Additional evidence of this is shown in the dimensional stability (antishrink efficiencies) of epoxide-treated wood.

Antishrink efficiencies were determined by water soaking treated and untreated samples for 7 days and measuring the change in volume due to water adsorption. The highest antishrink

Antishrink Efficiency (R) of Southern Yellow Pine Blocks

Compound	Weight add-on %	R
Propylene oxide	20.4	51.3
	28.0	68.1
	33.8	66.2
	37.7	35.4
	51.1	25.2
Epichlorohydrin	24.5	68.8
Butylene oxide	21.1	68.8

efficiencies (R) are in the range of 21-28% weight add-on. Above this level, the R values start to drop off which may mean the polymer loadings are so high they have broken the cell wall and allow the wood to superswell above the green volume.

In the epichlorohydrin samples, the chlorine was confirmed to be in the cell wall by energy dispersive analysis of X-ray spectra generated in the scanning electron microscope. The greatest percentage of chlorine was in the S_2 layer of the cell wall which is the thickest cell wall component and contains the most cellulose. Electron micrographs also showed no polymer in the lumen, but did show changes in the nature of the cell wall.

These findings of retention of chlorine in the reaction of epichlorohydrin under alkaline conditions contradict somewhat the literature on the mechanism of the reaction. The reaction as

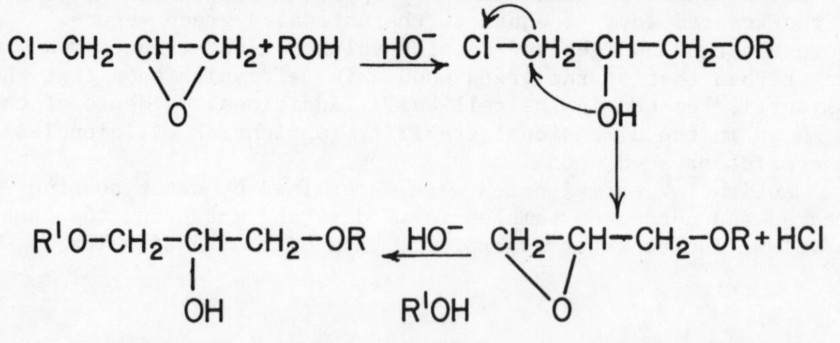

shown is reported to take place under strongly alkaline conditions (NaOH). The chlorohydrin undergoing internal reaction to give a new epoxide and HCl split out. The new epoxide would then be available for crosslinking. This reaction does not take place when the reaction is catalyzed by triethylamine. There is no loss of chlorine during the reaction and, if HCl was formed, there would be a drop in the pH after reaction. No pH drop was observed.

The effectiveness of these epoxide treatments as decay retardants was determined by soil-block tests using two brown-rot fungi. The brown rotting fungi are those which preferentially attack cellulose in the wood leaving the lignin alone. Treated and untreated southern yellow pine blocks were placed in test with the fungus *Lentinus lepideus* and separately with *Lenzites trabea*. Samples were removed at 6 and 12 weeks, and the extent of decay was determined by ovendry weight loss. The sample blocks from the reduction in water swelling test (7 days leaching) were also put in test to determine the changes in decay resistance as affected by leaching.

Soil-Block Tests on Propylene Oxide Treated Southern Yellow Pine Inoculated with *Lentinus lepideus*

Percent weight add-on	Percent weight loss in weeks	
	(6)	(12)
0	24.3	44.2
5.1	8.1	17.5
24.0	3.2	4.8
36.6	2.6	4.6
44.5	3.4	7.3
50.9	3.7	5.3

A weight loss after 12 weeks of less than 5% is regarded as a positive result.

Propylene and butylene oxides and epichlorohydrin all show good decay resistance to *Lentinus lepideus* at levels of about 23% and above. For southern yellow pine, *Lenzites trabea* is a much more severe decay fungus.

Soil-Block Tests on Treated Southern Yellow Pine Inoculated with *Lenzites trabea*

Sample	Nonleached		Leached	
	Percent weight loss after			
	6 weeks	12 weeks	6 weeks	12 weeks
Control	44.6	62.9	44.9	68.7
Propylene oxide, 20%	12.9	40.0	26.7	38.6
24	10.3	35.5	17.3	50.4
37	8.4	28.7	14.2	23.6
50	6.5	25.2	12.7	25.0
Epichlorohydrin, 17%	4.9	7.2	6.2	9.7
25	2.6	5.1	2.4	--
35	2.2	5.9	2.0	4.1
41	--	--	3.7	4.0
Butylene oxide, 7%	5.2	18.8	7.0	18.8
14	2.9	12.4	1.8	11.9
23	3.2	3.8	2.7	2.0

and the propylene oxide treatment does not hold up. For *trabea*, butylene oxide and epichlorohydrin give good decay resistance at levels above 22%.

In conclusion, the data from this work show that propylene oxide, butylene oxide, and epichlorohydrin treatments give good dimensional stability to water swelling at precent weight add-ons of approximately 25%. At these same levels of chemical substitution, two of the treatments show good rot resistance. These treatments may find applications in products such as window units and millwork in which resistance to swelling from water is as important as rot resistance.

Tests are now in progress to determine strength loss, if any, in the treated wood as well as studies on weathering, gluing, paintability, and burning characteristics.

9

Alkaline Degradation of a Nonreducing Cellulose. Model: 1,5-Anhydro-cellobiitol

RALPH E. BRANDON, LELAND R. SCHROEDER, and DONALD C. JOHNSON

The Institute of Paper Chemistry, Appleton, Wis. 54911

Abstract

Degradations of 1,5-anhydro-cellobiitol at 160-180°C in oxygen-free, 0.5-2.5N NaOH involve cleavage of both the glycosyl-oxygen bond (80-90%) and the oxygen-aglycon bond (10-20%). Cleavage of the oxygen-aglycon bond yields 1,5:3,6-dianhydro-D-galactitol (50-100%) and unidentified products (0-50%) from the aglycon, and is believed to occur by an S_N1 mechanism. Cleavage of the glycosyl-oxygen bond yields 1,5-anhydro-D-glucitol (100%) from the aglycon. The reactive intermediate, 1,6-anhydro-β-D-glucopyranose, is formed from the glycosyl moiety in *ca.* 35% of the cleavages of the glycosyl-oxygen bond and, hence, its formation is not as significant as is usually presumed. Glycosyl-oxygen bond cleavage does not appear to occur by a single mechanism and is probably governed by both $S_N1cB(2')$ and S_N1 mechanisms. In contrast to degradations of 1,5-anhydro-cellobiitol, degradations of 1,5-anhydro-2,3,6-tri-*O*-methyl-cellobiitol form *ca.* 65% 1,6-anhydro-β-D-glucopyranose from glycosyl-oxygen bond cleavage. The implications of these results with respect to alkaline cleavage of glycosidic bonds in cellulose are discussed.

Introduction

High-temperature alkaline processes involving cellulosic materials can result in a significant loss of weight and a drastic decrease in the degree of polymerization of the cellulose (1-3). The weight loss has been attributed primarily to endwise degradation ("peeling") of the polysaccharide while the drastic reduction in the degree of polymerization has been attributed primarily to random cleavage of glycosidic bonds (1).

While anaerobic alkaline degradations of aryl glycosides (4, 5) and alkyl glycosides (5) have been investigated extensively, and mechanisms have been proposed for these reactions, the mechanism of alkaline cleavage of glycosidic linkages joining monosaccharide units in oligo- or polysaccharides has received little attention. It is generally assumed (2, 3, 6) for cellulose that alkaline cleavage of the β-1,4-glycosidic linkages occurs by a neighboring group mechanism in which the aglycon is displaced by the conjugate base of the trans-2-hydroxyl group. The mechanism is analogous to the mechanism proposed by McCloskey and Coleman (7) to account for the alkaline lability of aryl trans-1,2-glycopyranosides, and which has subsequently been extrapolated, albeit questionably (5), to alkyl glycopyranosides. Best and Green (8) concluded that the data for the alkaline degradation of methyl β-cellobioside was consistent with such a mechanism, and a more recent kinetic analysis of these data by Lai (9) presumably strengthens this conclusion.

In this paper we report the results of a study of the mechanism of degradation of a nonreducing cellulose model; 1,5-anhydro-4-O-(β-D-glucopyranosyl)-D-glucitol, I (1,5-anhydro-cellobiitol); at 160-180°C in aqueous, oxygen-free sodium hydroxide (0.5-2.5N). Auxiliary studies of degradations of a partially-methylated derivative of I, 1,5-anhydro-4-O-(β-D-glucopyranosyl)-2,3,6-tri-O-methyl-D-glucitol (1,5-anhydro-2,3,6-tri-O-methyl-cellobiitol) are also reported.

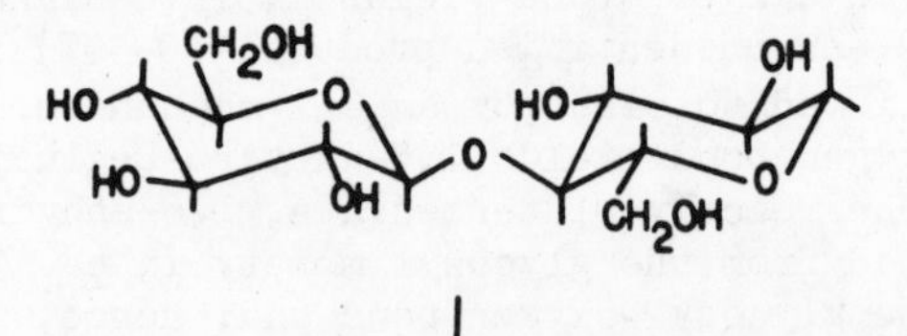

Results

Product Analyses. The product distribution for alkaline degradations of I depended on the reaction conditions. Stable, neutral products identified were 1,5-anhydro-D-glucitol (II, 80-90%), 1,5:3,6-dianhydro-D-galactitol (III, 8-11%), 1,5-anhydro-D-gulitol (IV, <0.5%), and 1,5-anhydro-D-galactitol (V, trace).

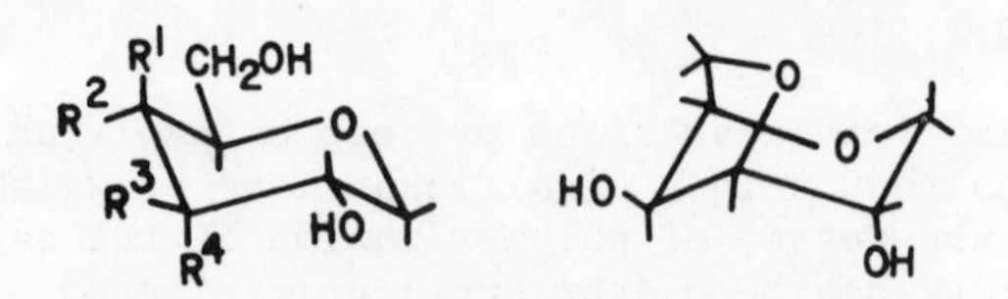

Compound II was identified by gas-liquid partition chromatography (g.l.c.) and paper chromatography (p.c.) and was demonstrated to be stable (<1% reaction) for a minimum of seven half-lives of I under the conditions employed (10). Compound III was initially identified by g.l.c.-mass spectrometry. In addition, III was isolated from a reaction mixture by column chromatographic techniques and subsequently identified by comparison of its n.m.r. and i.r. spectra, and g.l.c. and p.c. mobilities with authentic III. The stability of III to the reaction conditions was evident from the fact that the ratio of III to II in product mixtures did not change at long reaction times. Compounds IV and V were identified by g.l.c. and p.c. in column chromatographic-enriched fractions of a reaction mixture.

1,6-Anhydro-β-D-glucopyranose (VI) was identified as a reactive, intermediate product by g.l.c. and p.c. techniques.

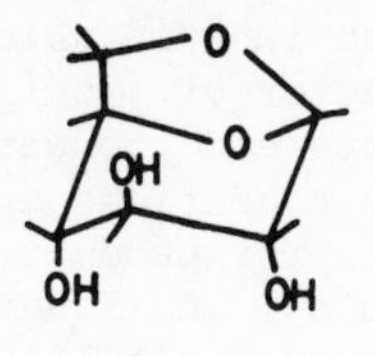

VI

Under some reaction conditions a difference (up to 12%) was observed between the moles of identified products formed from the aglycon (1,5-anhydro-D-glucitol moiety) of I and the moles of I reacted. This deficit is presumed to result from formation of either ionic or fragmentation products from the aglycon during degradation of I. These types of products would not be detected by the g.l.c. procedures employed. For purposes of further discussion, the mass deficit is referred to as the unidentified products (U).

Kinetic Analysis. The degradation of I was followed by quantitative g.l.c. analysis of deionized, acetylated samples of the reaction.

Since the hydroxide ion concentration was large, relative to the concentration of I, the disappearance of I and the appearance of stable products (II and III)[1] followed pseudo-parallel-first-order kinetics [Equations (1), (2), and (3); Fig. 1].[2]

[1] 1,5-Anhydro-D-gulitol (IV) and 1,5-anhydro-D-galactitol (V) are also stable products but their concentrations were too low to be measured routinely.

[2] Frost and Pearson (11) have described the mathematics of parallel-first-order kinetics. The equations given here are simplifications based on the fact that the term $k_i/\Sigma k_i$ is the mole fraction of the products accounted for by product i.

$$\ln(X_{r,t}) = -k_r t \quad (1)$$

$$\ln(X_{i,\infty} - X_{i,t}) = -k_r t + \ln(X_{i,\infty}) \quad (2)$$

$$k_i = k_r X_{i,\infty} \quad (3)$$

where $X_{r,t}$ is the mole fraction of reactant at time t, $X_{i,t}$ is the mole fraction of product i at time t, $X_{i,\infty}$ is $X_{i,t}$ at completion (the relative proportion of product i formed), k_r is the first-order specific rate constant for reactant disappearance ($k_r = \Sigma k_i$), and k_i is the first-order specific rate constant for formation of product i.

The degradation of I was also demonstrated to have a first-order dependence on the concentration of I by the half-life method (11) for two analogous reactions in which the initial concentration of I differed by a factor of two (10).

Specific rate constants and product distributions for degradations of I under various conditions are reported in Table 1. The specific rate constants for product formation can be calculated from the data in Table 1 and Equation (3). The analytical procedures provided good reproducibility; triplicate determinations of k_r for I at 170°C and 2.5N NaOH were within 1.5% of the mean value (10).

The specific rate constants (k_f) for the formation of 1,6-anhydro-β-D-glucopyranose (VI), necessary to calculate the mole fractions of VI formed in degradations of I ($X_{VI,\infty} = k_f/k_r$, Table 1)[3], were calculated (least squares) from the linear relationship [Equation (4), Fig. 2] describing the concentration of VI as a function of time (10).

$$L - L_0 \exp(-k_d t) = k_f R_0 (\exp(-k_r t) - \exp(-k_d t))/(k_d - k_r) \quad (4)$$

where L is the concentration of VI at time t, L_0 is the initial concentration of VI[4], R_0 is the initial concentration of reactant, k_r is the first-order rate constant for degradation of the reactant, k_f is the first-order rate constant for formation of VI, and k_d is the first-order rate constant for the degradation of VI.

Use of Equation (4) necessitated independent determination of the specific rate constant for degradation of VI (k_d) for each set of reaction conditions for which $X_{VI,\infty}$ was desired (Table 1).

Degradation of 1,5-Anhydro-2,3,6-tri-O-methyl-cellobiitol (VII). Degradation of VII in 0.5 and 2.5N NaOH at 2.5M ionic

[3]Since VI degrades in the systems studied, $X_{VI,\infty}$ is actually zero. Conceptually, however, $X_{VI,\infty}$ represents the mole fraction of I which has degraded via VI without reference to the subsequent fate of VI.

[4]Products could potentially be present at zero time which was chosen to be some short time (<1% reaction) after the reactor system had reached the desired temperature.

TABLE 1

RATE CONSTANTS AND PRODUCT MOLE FRACTIONS FOR DEGRADATION OF 1,5-ANHYDRO-CELLOBIITOL (I) AND 1,5-ANHYDRO-2,3,6-TRI-O-METHYL-CELLOBIITOL (VII)

No.	Reactant	Temp., °C	NaOH, M	NaOTs,[a] M	NaI, M	$10^6 k_r$, sec^{-1}	Product Mole Fraction ($X_{i,\infty}$)[b] II	III	VIII	U[c]	VI	$10^6 k_d$, sec^{-1}
1	I	170	2.5	--	--	7.89[d]	0.925	0.100	--	0.0	0.33	95
2	I	170	1.5	1.0	--	6.74	0.924	0.105	--	0.0	0.33	71
3	I	170	1.0	1.5	--	5.84	0.856	0.098	--	0.046	0.31	52
4	I	170	1.0	--	--	6.24	0.873	0.096	--	0.031	0.31	62
5	I	170	0.5	2.0	--	4.38	0.801	0.091	--	0.108	0.26	22
6	I	170	0.5	--	--	4.69	0.887	0.083	--	0.030	0.30	34
7	I	170	0.5	--	2.0	3.94	0.787	0.096	--	0.117	--	--
8	I	180	2.5	--	--	21.4	0.890	0.103	--	0.008	--	--
9	I	160	2.5	--	--	3.21	0.885	0.081	--	0.034	--	--
10	I	180	0.5	2.0	--	11.6	0.838	0.102	--	0.060	--	--
11	I	160	0.5	2.0	--	1.59	0.874	0.088	--	0.038	--	--
12	VII	170	2.5	--	--	8.49	--	--	0.875	0.125	0.64	95
13	VII	170	0.5	2.0	--	3.10	--	--	0.829	0.171	0.53	22

[a]Sodium p-toluenesulfonate; [b]1,5-Anhydro-D-glucitol (II), 1,5:3,6-dianhydro-D-galactitol (III), 1,6-anhydro-β-D-glucopyranose (VI), 1,5-anhydro-2,3,6-tri-O-methyl-D-glucitol (VIII), and unidentified products (U); [c]Unidentified products (U) are determined from a mass balance (see text), negative values are reported as 0.0; [d]Mean value of three determinations.

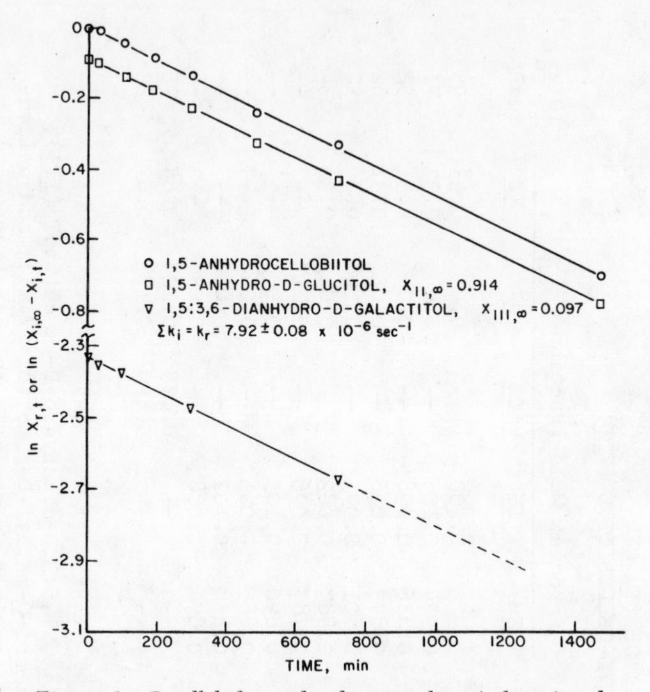

Figure 1. Parallel first-order kinetic plot of data for the degradation of 1,5-anhydro-cellobiitol in 2.5N NaOH at 170°C

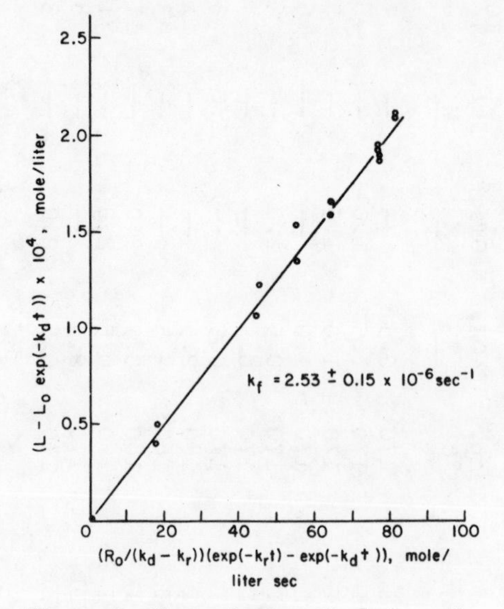

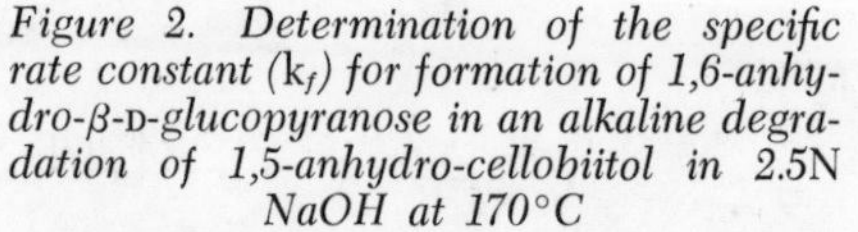

Figure 2. Determination of the specific rate constant (k_f) for formation of 1,6-anhydro-β-D-glucopyranose in an alkaline degradation of 1,5-anhydro-cellobiitol in 2.5N NaOH at 170°C

strength (Reactions 12 and 13, Table 1) yielded 1,5-anhydro-2,3,6-tri-O-methyl-D-glucitol (VIII) (83-88%) and unidentified products

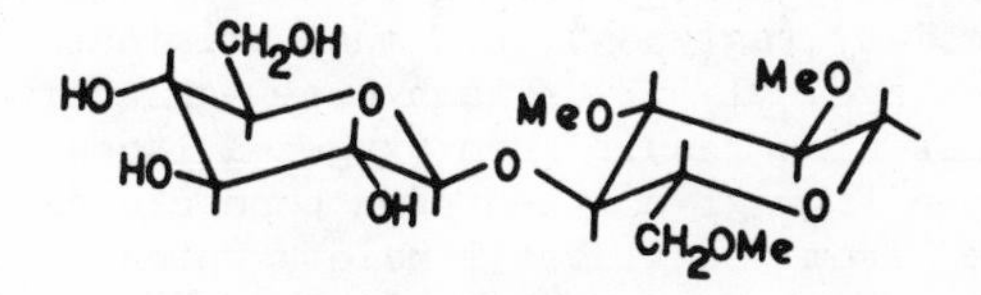

VII

(U)(12-17%). As discussed previously for degradations of I, U, indicated by a deficit in the mole balance of anhydroalditol moiety, is presumed to result from formation of ionic or fragmentation products from the aglycon.

Degradations of VII yielded approximately twice as much 1,6-anhydro-β-D-glucopyranose (VI) as degradations of I.

At 2.5M ionic strength, the reactivity of VII was less (29%) than that of I at 0.5N NaOH, but slightly greater (8%) at 2.5N NaOH.

Discussion

Point of Cleavage in the Glycosidic Linkage. Degradation of I results primarily from cleavage of the glycosyl-oxygen bond (A-A') of the glycosidic linkage, but cleavage of the oxygen-aglycon bond (B-B') also occurs to a significant extent.

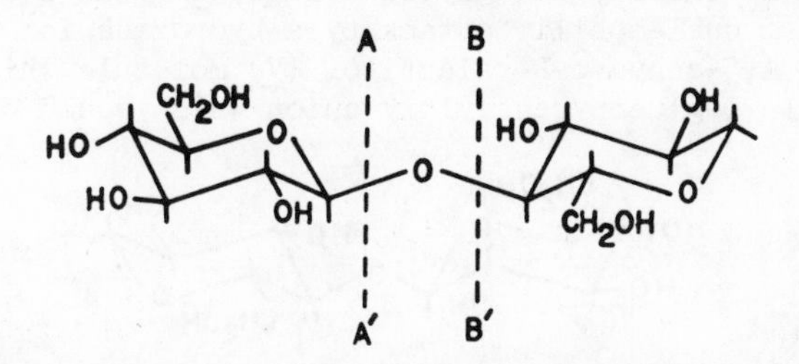

Cleavage of the glycosyl-oxygen bonds (A-A') in I and VII was considered to be the most probable mode of degradation since it was concluded (8) from isotope studies that methyl β-cellobioside degrades in this manner. In addition, glycosyl-oxygen bond cleavage must occur to the extent that VI is formed in these systems. This type of bond cleavage is also strongly suggested by the fact that the major products from I and VII are II and VIII, respectively.

Cleavage of the glycosyl-oxygen bond would yield only the conjugate base of 1,5-anhydro-D-glucitol (II) from the aglycon. Thus, since II was demonstrated to be stable under degradation conditions, the lack of a mole balance between II formed and I degraded leads to the conclusion that cleavage of the oxygen-aglycon bond must occur. Analogous reasoning indicates that oxygen-aglycon bond cleavage also occurs in degradations of VII. In addition, the unidentified products (U) must not only be derived from the anhydroalditol moiety (aglycon), but must originate from oxygen-aglycon bond cleavage. 1,5:3,6-Dianhydro-D-galactitol (III) formed from I must also result from oxygen-aglycon bond cleavage. If III were formed from glycosyl-oxygen bond cleavage, it would have to be formed from the glucosyl moiety which is unlikely, and it would also be expected to be a product of VII, which it is not.

Mechanism of Oxygen-aglycon Bond Cleavage. Characterization of oxygen-aglycon bond cleavage would be greatly complicated if a significant amount of 1,5-anhydro-D-glucitol (II) were formed during the process. However, this is considered to be very unlikely since cleavage of the interglycose linkage of methyl β-cellobioside under comparable conditions in $^{18}OH^-$ did not yield methyl β-D-glucopyranoside (analogous to II in the present system) containing detectable enrichment of ^{18}O (8). Thus, the products of oxygen-aglycon bond cleavage are primarily 1,5:3,6-dianhydro-D-galactitol (III) and the unidentified products (U). The specific rate constant for oxygen-aglycon bond cleavage, k_{OA}, can therefore be calculated from either Equation (5) or (6) in conjunction with the data in Table 1 and Equation (3).

$$k_{OA} = k_{III} + k_{U} \quad (5)$$

$$k_{OA} = k_{r} - k_{II} \quad (6)$$

Cleavage of the oxygen-aglycon bond of I by an S_N2 mechanism would involve a nucleophilic attack by a hydroxide ion at C-4 to form a stable 1,5-anhydro-D-galactitol (V) molecule while displacing an unstable D-glucopyranosyloxy anion which would degrade

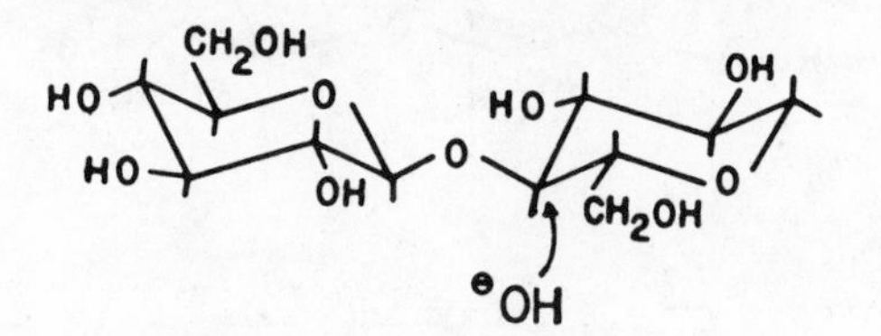

rapidly (12) to acidic products. Since only a trace amount of V was observed in the reaction products, an S_N2 mechanism is not an important route for cleavage of the oxygen-aglycon bond.

Typically, intramolecular displacement of the leaving group by the conjugate base of a properly oriented hydroxyl group has been proposed to be the rate-determining step in alkaline degradations of glycosides (5, 7-9, 13-15). The most logical mechanism of this type for oxygen-aglycon bond cleavage of I would be an $S_N1cB(3)$ mechanism. For the $S_N1cB(3)$ mechanism to occur, the 1,5-anhydro-D-glucitol moiety (aglycon) must be in the $^1C_4$ conformation. The conjugate base (cB) of the C-3 hydroxyl group could then potentially displace the D-glucopyranosyloxy anion from C-4 with concomitant formation of 1,5:3,4-dianhydro-D-galactitol which can, as will be described later, readily form the dominant product associated with oxygen-aglycon bond cleavage, 1,5:3,6-dianhydro-D-galactitol (III).

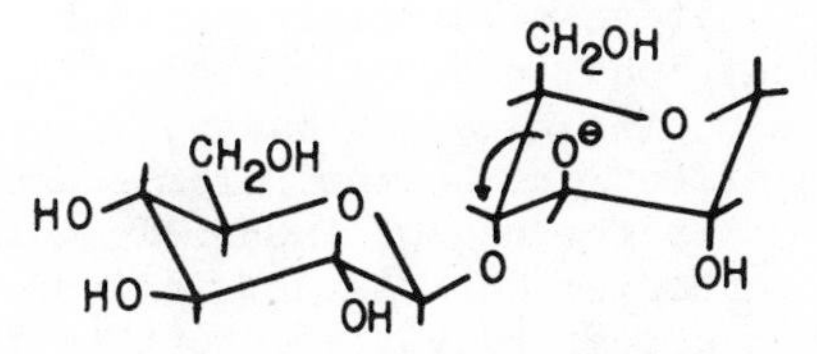

However, if an $S_N1cB(3)$ mechanism were important in oxygen-aglycon bond cleavage, blocking the C-3 oxygen with a methyl group should effect a drastic decrease in the rate of cleavage (16, 17). This is not the case. The rate constants, k_{OA}, for VII are not drastically different from those for I (Table 2). Increasing the alkali concentration would be expected to increase k_{OA} if an $S_N1cB(3)$ mechanism were operative because the concentration of the conjugate base of OH-3 would increase. However, at constant ionic strength, k_{OA} for I is not changed drastically and tends to decrease rather than increase as the hydroxide ion concentration is increased (Table 2). In addition, increasing the ionic strength from 0.5N to 2.5N NaOH effected an increase of ca. 65% in k_{OA} for I (Table 2). This is definitely inconsistent with an $S_N1cB(3)$ mechanism in which ionic charge would be delocalized in the transition state, and which would therefore not be accelerated, but probably retarded, by an increase in ionic strength.

TABLE 2

SPECIFIC RATE CONSTANTS FOR OXYGEN-AGLYCON BOND CLEAVAGE AT 170°C

		$10^6 k_{OA}$, sec^{-1}	
NaOH, M	NaOTs,[a] M	I[b]	VII[c]
2.5	--	0.77[d]	1.06
1.5	1.0	0.69[d]	--
1.0	1.5	0.84	--
0.5	2.0	0.87	0.53
0.5	--	0.53	--

[a]Sodium p-toluenesulfonate; [b]1,5-Anhydrocellobiitol, k_{OA} calculated from Equation (6); [c]1,5-Anhydro-2,3,6-tri-O-methylcellobiitol, $k_{OA} = k_r - k_{VIII}$; [d]Normalized value of $X_{III,\infty}$ used in calculations.

An S_N1 mechanism (Fig. 3), in which the oxygen-aglycon bond undergoes heterolysis to form a β-D-glucopyranosyloxy anion (X) and a 1,5-anhydro-4-deoxy-D-*xylo*-hexitol-4-cation (XI) as the initial products, can satisfactorily account for the observed results. In the transition state of the heterolysis, charge develops and, hence, a positive salt effect would be expected since the salt provides an additional mode of charge stabilization. The magnitude of the salt effect is not large, but this may be the result of a leveling effect due to the polar aqueous medium and the salt effect inherent in the use of NaOH. Inductive stabilization of the developing carbonium ion (XI) by ionized hydroxyl groups in the aglycon does not appear to be important to the heterolysis since, as noted previously, etherification of these hydroxyl groups does not drastically change k_{OA}, and k_{OA} does not increase as the hydroxide ion concentration increases.

The anion (X) should degrade rapidly to acidic products (12). The carbonium ion (XI) could potentially undergo an intermolecular reaction with hydroxide ion or water, an intramolecular reaction with a suitably disposed hydroxyl group or its conjugate base, a rearrangement, or an elimination reaction.

Reaction of XI with hydroxide ion or water to form 1,5-anhydro-D-galactitol (V) and 1,5-anhydro-D-glucitol (II) apparently is not important. If such a reaction occurred, it would have to be at a faster rate than other reactions of XI and, hence, it would be expected that V would be the dominant product because of a shielding effect of the departing anion X. The fact that V was formed in only trace amounts indicates that this reaction of VI is of negligible importance.

Intramolecular nucleophilic attack at C-4 of XI by OH-3 or its conjugate base (Fig. 3) can initiate a series of reactions which culminate in the formation of III, a major product of oxygen-aglycon bond cleavage. The initial intermediate in the sequence would be 1,5:3,4-dianhydro-D-galactitol (IX). Base-catalyzed epoxide migration would subsequently yield 1,5:2,3-dianhydro-D-gulitol (XII) from IX. Intramolecular nucleophilic attack at C-3 of XII by the conjugate base of OH-6 to open the oxirane ring would ultimately yield III. Cleavage of the oxirane rings of the intermediates, IX and XII, by hydroxide ion should yield 1,5-anhydro-D-gulitol (IV) and 1,5-anhydro-D-iditol (XIII), respectively, as the dominant products (18). This type of reaction could account for the small amount of IV found in the products. The formation of XIII was neither confirmed nor refuted. Further support for this reaction sequence is derived from the fact that in refluxing NaOH (0.1*N*) methyl 3,4-anhydro-α-D-galactopyranoside was converted to methyl 3,6-anhydro-α-D-galactopyranoside (crystalline yield, 71%) and trace amounts of methyl α-D-gulopyranoside and methyl α-D-idopyranoside (19).

Other potential reactions of the cation XI include proton eliminations to form unsaturated compounds or rearrangements involving hydride shifts to form more stable species. Both types of carbonium ion reactions are well known (20) and for XI either type of reaction could account for the unidentified products (U) as illustrated by reactions which would involve H-3. Rearrangement of XI *via* a hydride shift from C-3 to C-4 with concerted loss of a proton from OH-3 would yield 1,5-anhydro-4-deoxy-D-*erythro*-3-hexulose (XIV, Fig. 3). In the alkaline system XIV could undergo a β-elimination to open the ring and subsequently form the 2,3-diulose which would produce acidic products. Compound XIV could also react, reversibly, to form isomeric 1,5-anhydro-4-deoxy-D-hexuloses (21). Elimination of H-3 from XI would result in the formation of the enol XV which is a tautomer of XIV (Fig. 3).

The distribution of products from oxygen-aglycon bond cleavage varies considerably with the reaction conditions (Table 1). At 2.5*M* ionic strength, approximately an equimolar mixture of III and U is obtained in 0.5*N* NaOH, while only III is obtained in 2.5*N* NaOH. This suggests that less ionization of OH-3 increases the average lifetime of the carbonium ion XI, thereby allowing rearrangement or elimination reactions to become more important relative to formation of the 3,4-anhydride IX. At 0.5*N* NaOH, increasing the ionic strength from 0.5*M* to 2.5*M* approximately doubled the proportion of oxygen-aglycon bond cleavage which generated U. This can be rationalized on the basis that the effect of an increase in ionic strength on the concentration of the C-3 oxyanion would be small relative to the additional stabilization afforded to the initial carbonium ion XI.

Mechanism of Glycosyl-oxygen Bond Cleavage. Glycosyl-oxygen bond cleavage in I is characterized by exclusive formation of 1,5-anhydro-D-glucitol (II) from the aglycon and partial formation of 1,6-anhydro-β-D-glucopyranose (VI) from the glucosyl moiety. Since VI can only be formed as a result of glycosyl-oxygen bond cleavage, the mole fraction of VI based on the overall degradation of I ($X_{VI,\infty}$, Table 1) could be misleading. Therefore, the $X_{VI,\infty}$ values have been converted to mole fractions based only on glycosyl-oxygen bond cleavage ($Y_{VI,\infty}$, Table 3).

As discussed previously, II is not formed in more than trace amounts from oxygen-aglycon bond cleavage. Thus, since II is the sole product from the aglycon in glycosyl-oxygen bond cleavage, the specific rate constant for the reaction, k_{GO}, (Table 3) can be calculated from Equation (7) in conjunction with the data in Table 1.

$$k_{GO} = k_{II} = k_r X_{II,\infty} \qquad (7)$$

In contrast to oxygen-aglycon bond cleavage, glycosyl-oxygen bond cleavage is very dependent on the base concentration, with k_{GO} increasing nonlinearly with increasing hydroxide ion concentration at constant ionic strength[5] (Table 3).

[5]Most studies of the effect of hydroxide ion concentration on degradations of glycosides have not been made at constant ionic strength and, hence, interpretation of the experimental results is complicated by an uncertainty of the magnitude of the inherent salt effects.

TABLE 3

SPECIFIC RATE CONSTANTS AND MOLE FRACTIONS OF 1,6-ANHYDRO-β-D-GLUCOPYRANOSE (VI) FORMED FOR GLYCOSYL-OXYGEN BOND CLEAVAGE AT 170°C

NaOH,	NaOTs,[a]	NaI,	$10^6 k_{GO}$, sec^{-1}		$Y_{VI,\infty}$	
M	M	M	I[b]	VII[c]	I[b]	VII[c]
2.5	--	--	7.12[d]	7.43	0.37[d]	0.73
1.5	1.0	--	6.05[d]	--	0.37[d]	--
1.0	1.5	--	5.00	--	0.36	--
0.5	2.0	--	3.51	2.57	0.32	0.64
0.5	--	--	4.16	--	0.34	--
0.5	--	2.0	3.10	--	--	--

[a]Sodium p-toluenesulfonate; [b]1,5-Anhydro-cellobiitol, k_{GO} calculated from Equation (7); [c]1,5-Anhydro-2,3,6-tri-O-methyl-cellobiitol, $k_{GO} = k_{VIII} = k_r X_{VIII,\infty}$; [d]A normalized value of $X_{II,\infty}$ was used in calculations.

Lai (9) has shown that, for an ideal solution, if cleavage of the glycosidic linkage is anchimerically assisted by a conjugate base of a hydroxyl group of the glycoside, the observed rate constant, k_{obs}, should be related to the hydroxide ion concentration by Equation (8), or its reciprocal, Equation (9).

$$k_{obs} = Kk[OH^-]/(1 + K[OH^-]) \quad (8)$$

$$1/k_{obs} = 1/k + (1/kK)(1/[OH^-]) \quad (9)$$

where K is the equilibrium constant for formation of the conjugate base of the appropriate hydroxyl group and k is the specific rate constant for conversion of the ionized glycoside to products. If the hydroxide ion concentration is varied without maintaining constant ionic strength, the activity coefficient of the hydroxide ion changes (22). Under these conditions, it has been suggested (17) that $[OH^-]$ in Equations (8) and (9) should be replaced by the function K_w/h_-, where K_w is the dissociation constant for water and h_- is related to the acidity function H_- (17, 23). If, however, the ionic strength is held constant while the base concentration is varied, as in the present study, use of Equations (8) and (9) should be appropriate.

A plot of $1/k_{GO}$ versus $1/[OH^-]$ is linear, but this does not of itself, as claimed by Lai (9) for similar systems, confirm that glycosyl-oxygen bond cleavage occurs with anchimeric assistance from the conjugate base of the C-2' hydroxyl group, i.e., an $S_N1cB(2')$ mechanism. Any mechanism in which the rate-determining step involves cleavage of the glycosidic bond of a conjugate base of I *via* a unimolecular process would give comparable results. An S_N2 mechanism in which nucleophilic attack by hydroxide ion occurs only on the un-ionized substrate would also yield a linear correlation between $1/k_{GO}$ and $1/[OH^-]$. However, the theoretical significance of the slope and the intercept for an S_N2 mechanism would be different from the S_N1cB mechanisms. In addition, a linear reciprocal plot of this nature does not preclude the possibility of multiple mechanisms, some of which may not obey the reciprocal relationship. This is illustrated by the fact that the specific rate constant for the overall degradation of I, k_r, also yields a linear reciprocal plot yet k_r, depending on the reaction conditions, includes at least 10-20% S_N1 character (oxygen-aglycon bond cleavage). Thus, kinetic analysis of this type must be applied judiciously.

An S_N2 mechanism in which nucleophilic attack by hydroxide ion occurs at C-1' of I to form II and D-glucopyranose is not a likely route for glycosyl-oxygen bond cleavage. Addition of a more nucleophilic species (I^- Table 3) than hydroxide ion[6] at

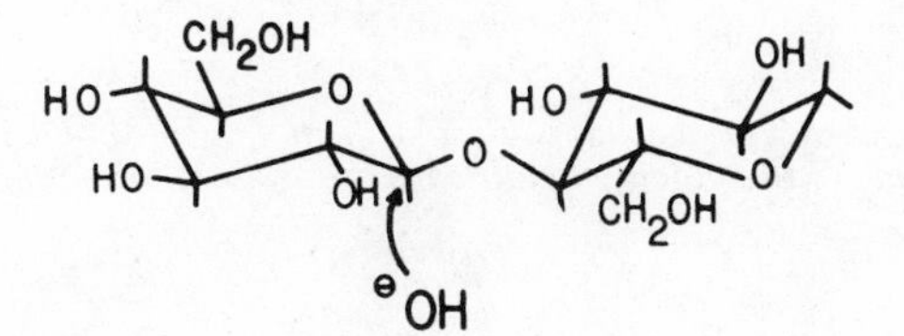

constant ionic strength and hydroxide ion concentration did not increase k_{GO} as would be expected if the reaction were governed by an S_N2 mechanism. In addition, since the initial product from the glucosyl moiety of I via an S_N2 mechanism would be D-glucopyranose, which would degrade rapidly to products other than its 1,6-anhydride (VI) (21), that portion of the reaction which generates VI ($Y_{VI,\infty}$, Table 3) cannot occur by an S_N2 mechanism. However, VI is not the major initial product of glycosyl-oxygen bond cleavage, as has been assumed in excluding the S_N2 mechanism for similar systems (9).

Since an S_N2 mechanism for glycosyl-oxygen bond cleavage is unlikely, the positive dependence of k_{GO} on the base concentration in conjunction with the negative salt effect (Table 3) suggests that an S_N1cB mechanism dominates the reaction. The most plausible mechanism of this type is the $S_N1cB(2')$ mechanism (Path A, Fig. 4) which is analogous to the mechanism proposed by McCloskey and Coleman (7) for the alkaline degradation of phenyl β-D-glucopyranoside. A rapid equilibrium between OH-2' and its conjugate base precedes the rate-determining step in which, with the glucopyranosyl moiety in the $^1C_4$ conformation, a nucleophilic attack by

[6]It is assumed from room temperature data (24) that iodide ion is a stronger nucleophile than hydroxide ion.

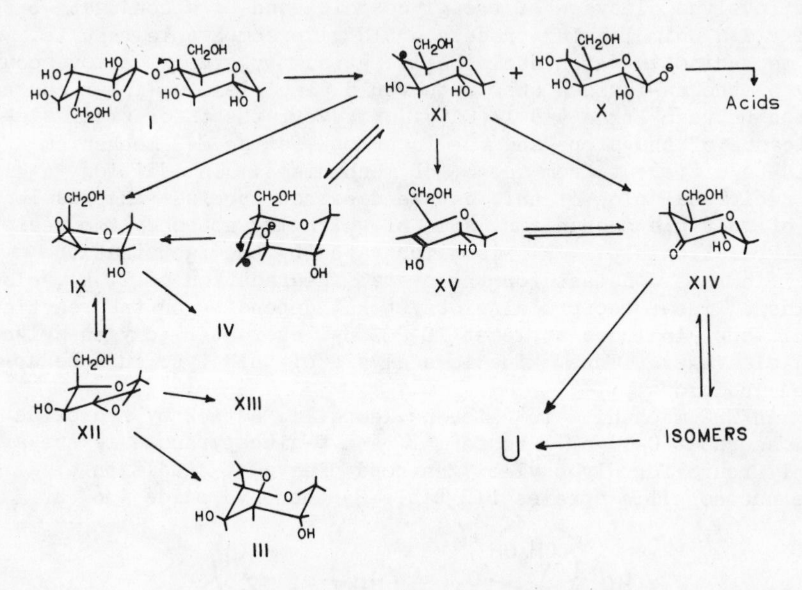

Figure 3. Proposed S_N1 mechanism for oxygen-aglycon bond cleavage with some of the potential pathways for product formation

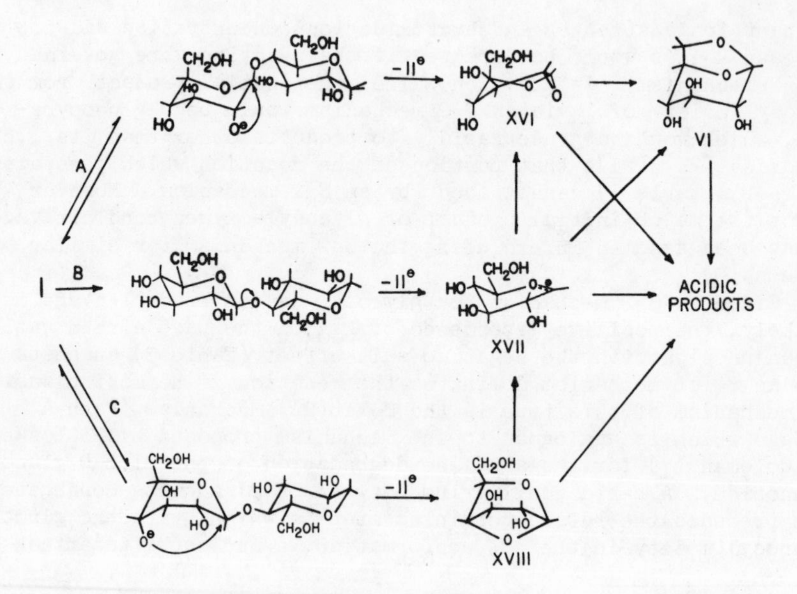

Figure 4. Possible mechanisms for glycosyl–oxygen bond cleavage. A: $S_N1cB(2')$. B: S_N1. C: $S_N1cB(4')$.

the C-2' oxyanion at C-1' displaces the conjugate base of II with concomitant formation of 1,2-anhydro-α-D-glucopyranose (XVI). The 1,2-anhydride XVI can subsequently yield VI by intramolecular nucleophilic attack at C-1 by the C-6 oxyanion ($^1C_4$ conformation) or react in several ways to yield degradation products via a reducing sugar.

The 1,6-anhydride VI can be isolated in high yield (88%) from the alkaline degradation of phenyl β-D-glucopyranoside under relatively mild conditions (100°C, 1.3N KOH) (7). Thus, the much lower "yield" of VI from glycosyl-oxygen bond cleavage of I (32-37%, Table 3) could indicate either that mechanisms other than an S_N1cB(2') are also operative or that the reactions which lead to degradation products from XVI become more significant relative to formation of VI as the temperature increases. The latter explanation, however, seems unrealistic in view of the relatively high "yield" of VI obtained from glycosyl-oxygen cleavage in VII (64-73%, Table 3). The only obvious difference between the reactants, I and VII, is that the aglycon hydroxyl groups of VII have been methylated; the glucosyl moiety, which must form XVI and subsequently VI, is the same for both compounds. Hence, the difference in the amount of VI formed from I and VII would appear to indicate at least a mixed mechanism for the reaction in I, and a shift toward more of the S_N1cB(2') mechanism for VII. The reason for such a mechanistic shift, however, is not known.

An S_N1cB(4') mechanism (Path C, Fig. 4), analogous to the mechanism proposed by Capon (5) for the alkaline degradation of phenyl β-D-mannopyranoside, is considered unlikely. Methylation of OH-2 of p-nitrophenyl β-D-xylopyranoside decreased the rate of its alkaline degradation by a factor of 1000 and effected a change in the reaction mechanism from S_N1cB(2) to bimolecular nucleophilic aromatic substitution rather than to S_N1cB(4) (17).

Thus, the most logical mechanism to operate in conjunction with the S_N1cB(2') mechanism for glycosyl-oxygen bond cleavage is the S_N1 mechanism (Path B, Fig. 4) in which heterolysis of the bond would initially yield the conjugate base of II and the resonance-stabilized D-glucopyranosyl cation (XVII). Reaction of the cation XVII with water or hydroxide ion would yield D-glucopyranose which would degrade. Intramolecular reactions of XVII involving OH-6 and OH-2, or their conjugate bases, would yield VI and XVI, respectively. As discussed previously, the dependence of k_{GO} on the base concentration does not preclude the possibility of the reaction having significant S_N1 character, particularly if bond heterolysis is aided, inductively, by ionization of hydroxyl groups in the D-glucopyranosyl moiety.

The lack of reference reactions for which the mechanisms are known makes it necessary to temper any conclusions drawn from apparent thermodynamic functions of activation for high temperature alkaline degradations of glycosides. However, the enthalpy of activation (ΔH^*) should be greater for the S_N1 than for the S_N1cB(2') mechanism since in the latter mechanism bond cleavage is being assisted by bond formation in the transition state. The entropy of activation (ΔS*) should also be greater for the S_N1 mechanism than for the $S_N1cB(2')$ mechanism since in the transition state of the latter reaction the nucleophilic substituent loses rotational freedom. In 2.5_ NaOH at 170°C, ΔH* and ΔS* for glycosyl-oxygen bond cleavage (37.1 kcal/mole and 1.0 e.u.) are lower than the corresponding functions for oxygen-aglycon bond cleavage (41.7 kcal/mole and 6.9 e.u.) which occurs by an S_N1 mechanism. However, since the differences are not very large, ca. 5 kcal/mole and 6 e.u., the S_N1 mechanism could be operating in conjunction with the $S_N1cB(2')$ mechanism.

Implications of the Results of this Study with Respect to the Alkaline Degradation of Cellulose. The random cleavage of glycosidic linkages in cellulose has generally been assumed to occur by the $S_N1cB(2)$ mechanism (2, 3, 6). This mechanism, however, requires that the pyranose ring assume the $^1C_4$ conformation prior to bond cleavage. In a long-chain polymer such as cellulose the "inertial" forces involved in rotating the substituents on C-1 and C-4 (glycosidic linkages) from the equatorial to the axial position should be considerably larger than those required in model compounds. In addition, the fact that the polymer chain is locked in a matrix (cellulose fiber) would further inhibit the change in conformation. This would seem to raise some fundamental questions as to the applicability of the $S_N1cB(2)$ mechanism to the alkaline cleavage of the cellulose polymer chain.

The S_N1 mechanisms do not require a change in the pyranose ring conformation prior to bond cleavage. Since the S_N1 mechanisms appear to be competitive with the $S_N1cB(2')$ mechanism in 1,5-anhydro-cellobiitol, the probability of these mechanisms being operative in the more restricted cellulose system would appear to be quite good.

Experimental

Analytical Methods. Melting points were determined on a Thomas-Hoover capillary apparatus which was calibrated against known compounds. Polarimetric measurements were made on a Perkin-Elmer 141MC polarimeter. I.r. spectra were determined on a Perkin-Elmer 621 grating spectrophotometer. N.m.r. spectra were determined on a Varian A-60A spectrometer at normal probe temperature using tetramethylsilane and 2,2-dimethyl-2-silapentane-5-sulfonate as the internal standards in $CDCl_3$ and D_2O solutions, respectively. Mass spectra were determined on a DuPont Instruments 21-491 spectrometer interfaced with a Varian Aerograph 1440-1 gas chromatograph.

T.l.c. was performed on plates coated with silica gel G (Brinkman Instruments). The components were located by spraying with sulfuric acid in methanol (1:5, vol) and subsequent charring.

P.c. was performed on Whatman No. 1 paper with ethyl acetate-pyridine-water (8:2:1, vol) as the developer. Chromatograms were developed for 35-40 hr and visualized with alkaline silver nitrate reagents (25).

A Varian Aerograph 1200-1 gas chromatograph equipped with a hydrogen flame ionization detector was used for g.l.c. Chromatographic response was recorded and integrated with a Honeywell Electronic 16 recorder equipped with a Disc integrator. All columns were arranged for on-column injection. Prepurified nitrogen (Matheson Gas Products) was used as the carrier gas. The conditions employed are listed below as: column type; N_2 flow, ml/min; column temp., °C; injector temp., °C; and detector temp., °C.

A: 10% SE-30 on 60-80 mesh AW-DMCS Chromosorb W (3 ft x 0.125 in. o.d.); 60; 140→212° at 6° min^{-1}, 212→310° at 15° min^{-1}; 275°; 350°.

B: Same as A except the column temperature was 140→310° at 4° min^{-1}.

C: 10% SE-30 on 60-80 mesh AW-DMCS Chromosorb W (3 ft x 0.125 in. o.d.); 20; 130→230° at 6° min^{-1}, 230→310° at 15° min^{-1}; 270°; 350°.

D: 5% SE-30 on 60-80 mesh AW-DMCS Chromosorb W (10 ft x 0.125 in. o.d.); 30; 180°; 250°; 300°.

Reagents. Reagent-grade, anhydrous sodium iodide was dried (105°C, 15 hr) and stored over NaOH pellets. Transfers and weighings were performed in a dry atmosphere.

Crystallization of commercial sodium p-toluenesulfonate (NaOTs) from ethanol containing 10-20% water gave platelets which contained less than 0.1% water (Karl Fischer Method). The purified material was stored over NaOH pellets. It was determined by g.l.c. that less than 0.001% of the major alkaline decomposition product of NaOTs, p-cresol, was present after subjecting 0.5M NaOTs with 2.0N NaOH to 170°C for 168 hr.

Hepta-O-acetyl-α-cellobiosyl Bromide (XIX). A slurry of α-cellobiose octaacetate (26) (132 g) in 1,2-dichloroethane (400 ml) was allowed to react with hydrogen bromide in acetic acid (30-32%, 135 ml) for 1.0 hr; diluted with $CHCl_3$ (800 ml); stirred with ice and water (2.7 l) for 0.5 hr; washed with water (1.0 l) saturated $NaHCO_3$ (1.0 l), and water (1.0 l); dried ($CaCl_2$); and concentrated in vacuo to ca. 450 ml. Addition of petroleum ether (b.p. 30-60°C) and refrigeration yielded crystalline XIX (105 g, 77%); m.p. 181-182.5°C (decomp.). Literature: m.p. 180°C (decomp.) (27).

1,5-Anhydro-cellobiitol (I). Compound XIX (14 g), 10% palladium-on-carbon catalyst (0.5 g), triethylamine (4 ml), and absolute ethyl acetate (150 ml) were hydrogenated (40-50 psig), with stirring, for 12-24 hr in a modified Parr bomb (28, 29). To monitor the reduction, samples were treated with silver nitrate (3%) in acetone-water (19:1, vol) to hydrolyze residual glycosyl bromide and then analyzed by t.l.c. (ethyl ether-pyridine; 20:1, vol). The reaction mixture was filtered, diluted with $CHCl_3$ (150 ml), washed with $NaHCO_3$ (2 x 300 ml) and water (300 ml), shaken with silver nitrate solution (3% in aqueous acetone, 5 ml), dried ($CaCl_2$), filtered (Celite), and concentrated in vacuo to a solid.

Two crystallizations from absolute ethanol gave hepta-O-acetyl-1,5-anhydro-cellobiitol (XX) (68%); m.p. 193.5-194°C, $[\alpha]_D^{24.5}$ + 4.1 ($CHCl_3$). Literature: m.p. 194°C, $[\alpha]_D^{25}$ + 4.6° ($CHCl_3$) (28).

Deacetylation of XX with sodium methoxide in methanol (30) and two crystallizations from 95% ethanol yielded I; m.p. 204.5-205.5°C, $[\alpha]_D^{24.5}$ 28.2° (H_2O). Literature: m.p. 172°C, $[\alpha]_D^{25}$ 29.5° (H_2O) (28); m.p. 204.5-205.5°C, $[\alpha]_D^{31}$ 28.6° (H_2O) (31).

1,5-Anhydro-D-glucitol (II). Tetra-O-acetyl-α-D-glucopyranosyl bromide (32) was reduced with lithium aluminum hydride in tetrahydrofuran (33). Crystallization of the product from absolute ethanol gave II (67%); m.p. 142-143°C, $[\alpha]_D^{24}$ 42.8° (H_2O). Literature: m.p. 142-143°C, $[\alpha]_D^{20}$ 42.8° (H_2O) (33).

1,5-Anhydro-D-galactitol (V). Tetra-O-acetyl-α-D-galactopyranosyl bromide (34) was reduced as described in the preparation of I. Isopropyl ether-pyridine (10:1, vol) was used for t.l.c. analyses. Crystallization of the product from absolute ethanol gave tetra-O-acetyl-1,5-anhydro-D-galactitol (XXI) (51%); m.p. 73-74°C, $[\alpha]_D^{25}$ 47.6° ($CHCl_3$). Literature: m.p. 75-76°C, $[\alpha]_D^{20}$ 49.1° ($CHCl_3$) (35); m.p. 103-104°C, $[\alpha]_D^{20}$ 47.9° ($CHCl_3$) (36).

Deacetylation of XXI with sodium methoxide in methanol (30) and crystallization from absolute ethanol yielded V; m.p. 112-113°C, $[\alpha]_D^{24.5}$ 76.5° (H_2O). Literature: m.p. 113-114°C, $[\alpha]_D^{20}$ 76.6° (H_2O) (35).

1,5:3,6-Dianhydro-D-galactitol (III). 1,5-Anhydro-2,3,4-tri-O-benzoyl-6-O-(p-toluenesulfonyl)-D-galactitol prepared from V was treated with sodium methoxide in methanol to yield III (37); m.p. 143-145°C (from EtOAc), $[\alpha]_D^{25}$ 40.0° (H_2O). Literature: m.p. 145-146°C, $[\alpha]_D^{20}$ 40.2° (H_2O) (37).

1,5-Anhydro-D-gulitol (IV). Compound XXI was treated with liquid hydrogen fluoride to produce a mixture of IV and V (36). The product mixture was refluxed with 0.5N NaOH overnight, deionized (Amberlite MB-3), and concentrated in vacuo for use as a reference material for p.c. and g.l.c. analyses.

1,5-Anhydro-2,3,6-tri-O-methyl-D-glucitol (VIII). 2,3,6-Tri-O-methyl-D-glucopyranose (10, 39) was acetylated with pyridine-acetic anhydride (38) using a modified work-up in which all aqueous phases were extensively back-extracted with $CHCl_3$. The sirupy, acetylated product was used to prepare crude 4-O-acetyl-2,3,6-tri-O-methyl-α-D-glucopyranosyl bromide as described for XIX. The bromide was reduced as described for I except for employing greater hydrogen pressure (70 psig), extensive $CHCl_3$ back-extraction of aqueous phases in the work-up procedure, and chloroform-ethyl acetate (2:1, vol) in t.l.c. analyses. Deacetylation (30) of the product mixture followed by purification on a silica gel (Grace Grade 950, 60-200 mesh) column with $CHCl_3$-MeOH (20:1, vol), vacuum distillation, and crystallization from isopropyl ether gave VIII; m.p. 32-32.5°C, $[\alpha]_D^{25}$ 53.8° (H_2O), $[\alpha]_{546}^{25}$ 63.5° (H_2O). (Found: C, 52.5; H, 8.9. $C_9H_{18}O_5$ requires: C, 52.4; H, 8.8%.) The n.m.r. spectrum (D_2O) of VIII had three methoxy singlets (δ 3.38, 3.48, and 3.60 ppm).

1,5-Anhydro-2,3,6-tri-O-methyl-cellobiitol (VII). Powdered Drierite (10 g), silver oxide (2 g), VIII (1.08 g), and absolute chloroform (45 ml) were rotated at a 45° angle in a light-protected, round-bottom flask for 0.5 hr. Tetra-O-acetyl-α-D-glucopyranosyl bromide (4 g) and iodine (0.05 g) were added to the mixture and mixing was continued. Additional glucosyl bromide (4 g) and silver oxide (2 g) were added to the reaction at 20 and 44 hr. To monitor the reaction, samples were treated with silver nitrate (3%) in acetone-water (19:1, vol) to hydrolyze residual glucosyl bromide, sodium chloride in acetone-water (19:1, vol) to precipitate excess silver ion, and analyzed by g.l.c. (Conditions B). After 54 hr, the reaction mixture was filtered and the residue was rinsed with $CHCl_3$ (100 ml). The combined filtrates were washed with $NaHCO_3$ (150 ml) and water (100 ml), dried ($CaCl_2$), and evaporated, *in vacuo*, to a thick sirup. The sirup was acetylated with pyridine-acetic anhydride (38) and separated into crude mono- and disaccharide fractions on a silica gel column (Grace Grade 950, 60-200 mesh, 275 g; 25 x 1000 mm) eluted with chloroform-ethyl acetate (2:1, vol). The disaccharide fraction was deacetylated (30) and fractionated on a silica gel column (100 g, 25 x 500 mm) with chloroform-methanol (7:1, vol) to give VII as the first disaccharide component eluted. Since VII could not be induced to crystallize, it was dried *in vacuo* to an amorphous solid (32% yield), $[\alpha]_D^{25}$ 15.3° (H_2O). (Found: C, 48.9; H, 7.6. $C_{15}H_{28}O_{10}$ requires: C, 48.9; H, 7.7%.)

Acid hydrolysis of VII gave VIII and α,β-D-glucose as determined by g.l.c. (Conditions C) for the per-O-trimethylsilyl ethers. The β-configuration of the glycosidic linkage of VII was confirmed by a doublet (H-1', δ 4.45 ppm, $J_{1',2'}$ 7.0 Hz) in its n.m.r. spectrum (D_2O) which was characteristic of an anomeric proton associated with a β-D-glucopyranosidic bond (40) and similar to the H-1' doublet of I (δ 4.51 ppm, $J_{1',2'}$ 7.0 Hz).

Acetylation of VII with acetic anhydride-pyridine (38) gave 1,5-anhydro-2',3',4',6'-tetra-O-acetyl-2,3,6-tri-O-methyl-cellobiitol; m.p. 80-80.5°C (from MeOH), $[\alpha]_D^{24}$ 15.7° ($CHCl_3$), $[\alpha]_{546}^{24}$ 18.6° ($CHCl_3$). (Found: C, 51.3; H, 6.8. $C_{23}H_{36}O_{14}$ requires: C, 51.5; H, 6.7%.)

N-Butyl β-D-glucopyranoside (XXII). Deacetylation (30) of n-butyl tetra-O-acetyl-β-D-glucopyranoside (41) gave XXII; m.p. 67-68°C (from EtOAc), $[\alpha]_D^{20}$ - 36.7° (H_2O). Literature: m.p. 66-67°C, $[\alpha]_D$ - 37.4° (H_2O) (42).

Cyclohexyl β-cellobioside (XXIII). Reaction of XIX with cyclohexanol (41) yielded cyclohexyl β-cellobioside heptaacetate (XXIV) (62%); m.p. 202-203.5°C (from EtOH), $[\alpha]_D^{25}$ - 25.7° ($CHCl_3$). (Found: C, 53.5; H, 6.5. $C_{32}H_{46}O_{18}$ requires: C, 53.5; H, 6.5%.) Deacetylation (30) of XXIV gave XXIII; m.p. 206.5-207.5°C (from EtOH), $[\alpha]_D^{25}$ - 26.3° (H_2O). (Found: C, 51.2; H, 7.6. $C_{18}H_{32}O_{11}$ requires: C, 50.9; H, 7.6%.) The n.m.r. spectrum (D_2O) of XXIII contained two anomeric proton doublets (δ 4.53 ppm, J 6.9 Hz and δ 4.59 ppm, J 7.5 Hz), both characteristic of β-glucopyranosidic bonds, thus confirming the β-configuration of the cyclohexoxy substituent.

Product Analysis. The presence of II, III, IV, V, and VI in reaction mixtures was demonstrated by g.l.c. analysis of the pertrimethylsilyl ethers (Conditions D) and p.c. The analysis and identification of III by g.l.c.-mass spectrometry is described in detail elsewhere (10). In addition, III was isolated from a large-scale degradation of I (ca. 8 g) in 2.5N NaOH at 170°C (71.5 hr). The reaction solution was deionized (Amberlite IR-120 and Amberlite MB-3) and concentrated in vacuo. The acetylated mixture was separated into crude mono- and disaccharide fractions on a silica gel column (Grace Grade 950, 60-200 mesh, 275 g; 25 x 1000 mm) using chloroform-ethyl acetate (2:1, vol) as the eluant. The trailing monosaccharide fractions were deacetylated and separated on silica gel (35 g, 12 x 500 mm) using chloroform-methanol (5:1, vol) to yield t.l.c. pure III which had i.r. and n.m.r. spectra virtually identical with those of an authentic sample. The n.m.r. spectrum of the acetylated III was also identical with that of known III diacetate.

Kinetic Analysis. A stock solution of 2.50N NaOH was prepared under a nitrogen atmosphere from carbon dioxide-free, triply-distilled water (43). The other NaOH solutions were prepared from the stock solution using similar water and, when appropriate, NaOTs and NaI. All alkaline solutions were stored under nitrogen in paraffin-lined bottles.

The reactor system, described in detail elsewhere (10), consisted of a type 316 stainless steel reactor (100-ml capacity) from which samples (ca. 1 ml) could be withdrawn, and an oil bath equipped with a Bronwell constant temperature circulator which could maintain the bath within 0.2°C of the desired temperature.

Oxygen was desorbed from the reactor by heating the disassembled reactor under vacuum (ca. 0.05 mm Hg, 105-110°C, 24-48 hr). The reactor was cooled under vacuum, and loaded (0.001 mole of reactant; 100 ml of NaOH solution) and assembled in a nitrogen atmosphere.

The reactor was connected to the sampling system and immersed in the oil bath. The initial sample for the arbitrary zero time was taken after the reactor had come to the desired temperature (<1% reaction) as indicated by an internal thermocouple. The sampling loop was purged just prior to sampling to insure sample uniformity. The actual sample size and the amounts of internal standard solutions added to the sample were determined gravimetrically. The amount of each internal standard solution XXII, ca. 0.003M and XXIII, ca. 0.007M) added to the sample was varied according to the amount of mono- and disaccharide estimated to be present.

The samples were deionized by passage over Amberlite MB-3 (6-8 ml, 15 x 150 mm column), concentrated in vacuo to dryness, and acetylated for 18 hr with acetic anhydride (0.25 ml) in pyridine (0.75 ml). Crushed, distilled-water ice (8 ml) was added to the sample, and the mixture was shaken mechanically for 0.5 hr. The mixture was extracted with $CHCl_3$ (2 x 5 ml). The $CHCl_3$ extracts were washed with N HCl (2 x 10 ml), H_2O (10 ml), saturated $NaHCO_3$ (10 ml), and H_2O (10 ml); dried (Na_2SO_4); and concentrated *in vacuo* to dryness. For reactions of VII, all aqueous solutions for the work-up were saturated with NaCl, and all aqueous phases were back-extracted with chloroform (5 ml). The dried sample was dissolved in $CHCl_3$ (*ca.* 0.2 ml) and analyzed by g.l.c. (Conditions A for I, Conditions C for VII). Response factors were determined by subjecting known mixtures of the necessary compounds to the analysis procedure.

Literature Cited

1. Corbett, W. M., and Richards, G. N., Svensk Papperstid. (1957) 60, 791.
2. Richards, G. N., Methods Carbohyd. Chem. (1963) 3, 154.
3. Lai, Y., and Sarkanen, K. V., Cellulose Chem. Technol. (1967) 1, 517.
4. Ballou, C. E., Adv. Carbohyd. Chem. (1954) 9, 59 and references cited therein.
5. Capon, B., Chem. Rev. (1969) 69, 407 and references cited therein.
6. Rydholm, S. A., "Pulping Processes," Interscience, New York, 1965.
7. McCloskey, C. M., and Coleman, G. H., J. Org. Chem. (1945) 10, 184.
8. Best, E. V., and Green, J. W., Tappi (1969) 52, 1321.
9. Lai, Y. Z., Carbohyd. Res. (1972) 24, 57.
10. Brandon, R. E., Doctoral Dissertation, The Institute of Paper Chemistry, Appleton, Wis., Jan., 1973.
11. Frost, A. A., and Pearson, R. G., "Kinetics and Mechanism," 2nd ed., John Wiley and Sons, Inc., New York, 1961.
12. Minor, J. L., Kihle, L. E., and Sanyer, N., Tappi (1969) 52, 2178.
13. Lindberg, B., Svensk Papperstid. (1956) 59, 531.
14. Brooks, R. D., and Thompson, N. S., Tappi (1966) 49, 362.
15. Robins, J. H., and Green, J. W., Tappi (1969) 52, 1346.
16. Gasman, R. G., and Johnson, D. C., J. Org. Chem. (1966) 31, 1830.
17. DeBruyne, C. K., Van Wijnendaele, F., and Carchon, H., Carbohyd. Res. (1974) 33, 75.
18. Williams, N. R., Adv. Carbohyd. Chem. (1970) 25, 109.
19. Buchanan, J. G., and Fletcher, R., J. Chem. Soc. (1965) 6316.
20. Bethell, D., and Gold, V., "Carbonium Ions. An Introduction," Academic Press, New York, 1967.
21. Pigman, W., and Anet, E. F. L. J. *In* Pigman and Horton's "The Carbohydrates," 2nd ed., Vol. IA, p. 165, Academic Press, New York, 1972.
22. Akerlof, G., and Kegeles, G., J.A.C.S. (1940) 62, 620.

23. Liberles, A., "Introduction to Theoretical Organic Chemistry," The Macmillan Company, New York, 1968.
24. Hine, J., "Physical Organic Chemistry," 2nd ed., McGraw-Hill Book Company, New York, 1962.
25. Trevelyan, W. E., Proctor, D. P., and Harrison, J. S., Nature (1950) 166, 444.
26. Braun, G., Org. Syntheses (1937) 17, 36.
27. Fischer, E., and Zemplen, G., Ber. (1910) 43, 2536.
28. Zervas, L., and Zioudrou, C., J. Chem. Soc. (1956) 214.
29. Gray, G. R., and Barker, R., J. Org. Chem. (1967) 32, 2764.
30. Thompson, A., Wolfrom, M. L., and Pacsu, E., Methods Carbohyd. Chem. (1963) 2, 216.
31. McCloskey, J. T., Doctoral Dissertation, The Institute of Paper Chemistry, Appleton, Wis., Jan., 1971.
32. Bates, F. J., "Polarimetry, Saccharimetry, and the Sugars," p. 500, U.S. Gov. Printing Office, Washington, D.C., 1942.
33. Ness, R. K., Fletcher, H. G., and Hudson, C. S., J.A.C.S. (1950) 72, 4547.
34. Capon, B., Collins, P. M., Levy, A. A., and Overend, W. G., J. Chem. Soc. (1964) 3242.
35. Fletcher, H. G., and Hudson, C. S., J.A.C.S. (1948) 70, 310.
36. Hedgley, E. J., and Fletcher, H. G., J.A.C.S. (1963) 85, 1615.
37. Fletcher, H. G., and Hudson, C. S., J.A.C.S. (1950) 72, 886.
38. Wolfrom, M. L., and Thompson, A., Methods Carbohyd. Chem. (1963) 2, 212.
39. Irvine, J. C., and Hirst, E. L., J. Chem. Soc. (1922) 121, 1213.
40. Minnikin, D. E., Carbohyd. Res. (1972) 23, 139.
41. Schroeder, L. R., and Green, J. W., J. Chem. Soc. (C) (1966) 530.
42. Kreider, L., and Friesen, E., J.A.C.S. (1942) 64, 1482.
43. Weissberger, A., ed., "Technique of Organic Chemistry," Vol. I, Part I, p. 256, Interscience, New York, 1949.

10

Nascent Polyethylene—Cellulose Composite

R. H. MARCHESSAULT, B. FISA, and J. F. REVOL

Department of Chemistry, Université de Montréal, Montréal, Québec, Canada

ABSTRACT

A technique has been developed which introduces new features to cellulose-synthetic polymer composites. Instead of melt coating or fibrous admixtures, polyethylene is directly synthesized within the matrix of a finished sheet or fibrous construction. For paper, most sheet properties are improved and due to the unique particulate polyethylene morphology, which is a characteristic of nascent polyethylene, brightness and opacity are markedly improved with less than 10% polyethylene "add on". Wet strength and water proofness are imparted while vapor permeability is maintained. The present state of development of a continuous process for treating paper, non-wovens and textiles is described.

INTRODUCTION

A new method for associating synthetic polymers and cellulose fibers is presently evolving: encapsulation. The operating principle consists in controlling and limiting the locus of polymerization to the fiber surface. In addition the polymer is prepared under conditions where polymerization and crystallization (or phase separation) of the growing polymer are nearly simultaneous so that a unique particulate morphology, termed nascent, results. In this paper we will limit our considerations to polyethylene but the principles of the method are applicable to many polymer systems.

By adsorbing a Ziegler-Natta catalyst at the external surface of a cellulose fiber, a layer of nascent polyethylene of controllable thickness can be synthesized thereon. Several means of surface activation are possible but in all cases one seeks to limit the locus of polymerization to the fiber topology and to texturize the polymer by controlled precipitation during polymerization.

When the method is applied for putting polyethylene on single fibers of cellulose, a hydrophobic surface is created and thermoplasticity becomes the interfiber bonding principle. On the other hand a fibrous construct such as paper or a non-woven leads to a complementary continuous matrix of polyethylene. Cellulose-cellulose interfiber bonds present in the original substrate are maintained and the continuous microporous nature of the polymer matrix can be demonstrated by solvent extraction of the cellulose, the resulting membrane being a perfect replica of the original inner surface of the fibrous construct.

Several examples of the above principles are reported:

a) N.L. Industries commercialized a battery separator some years ago based on single fiber encapsulates and subsequent sheet formation (1), (2), (3).

b) The "Centre Technique du Papier" of Grenoble have shown how the method can be applied to high yield mechanical pulps normally only used in particle boards. The encapsulated fibers could be heat-molded into water resistant yet porous sheets, especially suited to filter applications (4).

c) Two distinct processes have been used for direct encapsulation of paper. The entire fiber surface was made catalytic: β-$TiCl_3$ deposition in one case (5) and by VCl_3 in the other (4). The two methods illustrate the scope for devising new methods of catalyst deposition as we shall see in the following paragraphs.

An organometallic, such as aluminum triethyl ($AlEt_3$), acting on $TiCl_4$ yields (6) crystalline $TiCl_3$:

$$TiCl_4 + AlEt_3 \longrightarrow TiCl_3 + AlEt_2Cl + Et\cdot$$

The following process is used to immobilize the solid $TiCl_3$: the paper sheet is dipped in a solution of $TiCl_4$ in isopentane and subsequent evaporation of the solvent leaves a uniform deposit of liquid $TiCl_4$ in the paper matrix which is then plunged into a heptane solution of $AlEt_3$. The organometallic reduces the $TiCl_4$ *in situ* to crystalline $TiCl_3$.

Thermal decomposition of VCl_4 yields solid VCl_3

$$VCl_4 \xrightarrow{\Delta} VCl_3 + \tfrac{1}{2}Cl_2$$

In the Grenoble process paper sheets are contacted with VCl_4 vapors at a temperature of about 100°C which leads to a rather fine homogeneous deposit of VCl_3

which adheres to the surface of the cellulose fibers. Once the fibrous surface has been made catalytically active by one of the other method, it need only be contacted with ethylene in the presence of $AlEt_3$ and polymerization is instataneous leading to a heterogeneous but continous deposit of polyethylene.

MORPHOLOGY OF THE POLYETHYLENE ENCAPSULANT

To appreciate the properties of the encapsulated paper we consider the nature of nascent polyethylene. A recent review (7) surveys the numerous observations concerning this morphology which have appeared in recent polymer literature.

The word nascent generally refers to the reactive state of freshly generated atomic species. In its structural sense the nascent state is a concept used by biologists in referring to the initial conformation of a protein as it emerges from the active sites of the ribosomes. For synthetic polymers which are synthesized in a non-solvant "nascent" implies a simultaneous polymerization and crystallization (or precipitation). This is a "once in a lifetime" event for the macromolecule which under these circumstances develops a unique morphology. Whenever a supported catalyst is used, this morphology is best described as a pearl necklace assembly of submicron polyethylene particles. There is a one to one correlation between catalyst texture and the porosity and coarseness of the polymer continuous matrix; thus control of the particle size in the catalyst deposit leads to texture control in the encapsulant: a very fine catalyst deposit gives a dense uniform encapsulant while a catalyst made of large irregular shaped particle aggregates leads to a rough and porous membrane (8).

The three electron micrographs in Fig. 1 are meant to illustrate the foregoing description; they correspond successively to:

a) Ziegler-Natta catalyst on a glass substrate.
b) The first stages of polymerization.
c) After several minutes of polymerization.

The catalyst particles whose average dimensions are $\leqslant 0.1\ \mu$, are of ill-defined shape but are uniformly distributed on the substrate. After a few seconds of polymerization a significant quantity of polymer is visible around each of the catalyst particles. Suprisingly the dimensions of this initial particulate growth is quite uniform (9) and this carries through to the morphology of the finished product (Fig. 1c) where the particle size has reached a value of $\sim 0.5 - 0.6\ \mu$. This is in the desirable size range for organic pigments and provides some interesting enhancement of the optical properties of paper as we shall see.

The texture in Fig. 1c was obtained *via* the β-$TiCl_3$ catalyst as described above. When the quantity of polyethylene "add on" is less than 5% one still sees the fiber structure of the substrate (cf. Fig. 2). However as the add on goes beyond 10% the fiber matrix is lost and only the gossamer, nascent polyethylene is visible. At higher magnification the continuity and basic porosity of the structure is clearly visible (cf. Fig. 3).

The VCl_3 catalyst prepared as described above leads to a much finer texture and more homogeneous deposit of polyethylene as can be seen in Fig. 4. Even at high add-ons fiber shapes are clearly discernible. At higher magnification (Fig. 5) one sees multiple voids and fibrillation effects causes by the stresses created due to simultaneous polymerization and crystallization. Electron microscopic studies on the texture of the catalyst generated by thermal decomposition of VCl_4 have clearly shown that the finer particles in Fig. 4 are related to much finer particles of VCl_3 compared to β-$TiCl_3$. This correlation of nascent texture with that of the catalyst is an important aspect of nascent morphology and offers the prospect of being able to tailor the most desirable polymer structure for enhancing a given physical property of paper.

PHYSICAL PROPERTIES OF ENCAPSULATED PAPER

External appearance and "hand" of encapsulated paper depends markedly on the percent add on. Above ten percent the "hand" is warmer and optical effects become noticeable because of the finely divided nature of the polyethylene.

Even with a few percent of added polyethylene the sheet surface is non-wetting however the porosity of the surface remains as can be shown by the capillary sorption of hydrocarbon liquids.

Although the contact angle on the encapsulated paper surface is almost the same as for polyethylene film, it nevertheless is possible to rub water into the sheet. This is due to the microporosity which provides channels for movement of liquid water as well as for vapor transmission. For an unbleached Kraft paper which had been encapsulated using the VCl_3 route the water vapor permeability results shown in Table I were recorded (4).

The greater part of our physical measurements were recorded using the β-$TiCl_3$ method of encapsulation on Scheleicher and Schuell filter paper No. 595.

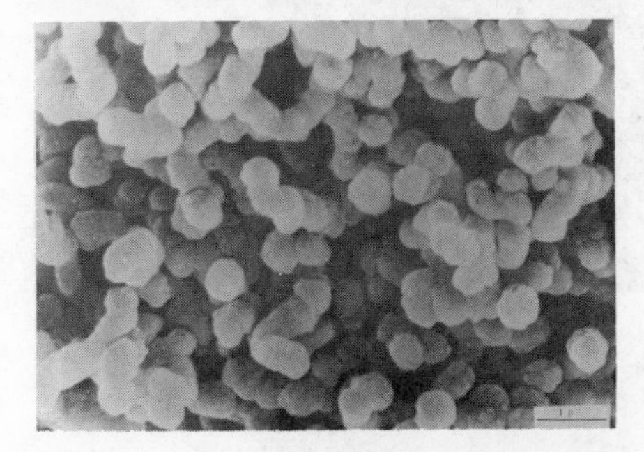

Figure 1a (top left). Transmission electron micrograph of carbon–platinum replica of glass surface covered with $\beta TiCl_3$. The latter was obtained by a plasma process acting on gaseous $TiCl_4$. Figure 1b (top right). Transmission electron micrograph of carbon–platinum replica of the same sample as in Figure 1a after several seconds of ethylene polymerization using heptane as medium and $AlEt_3$ as cocatalyst. Figure 1c (bottom left). Scanning electron micrograph of the sample in Figure 1a after 5 min of ethylene polymerization.

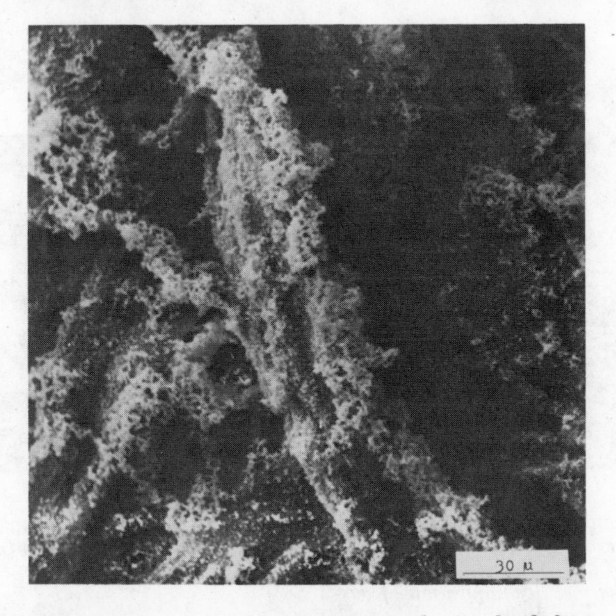

Figure 2. Scanning electron micrograph of polyethylene encapsulated paper surface with 5% "add on." The catalyst system was ($TiCl_4$ + $AlEt_3$) and aluminum triethyl as cocatalyst.

Figure 3. Scanning electron micrograph of the same sample as in Figure 2 but at higher magnification

Figure 4. Scanning electron micrograph of polyethylene encapsulated paper surface with 30% "add on." The catalyst was VCl_3 obtained by thermal decomposition of gaseous VCl_4.

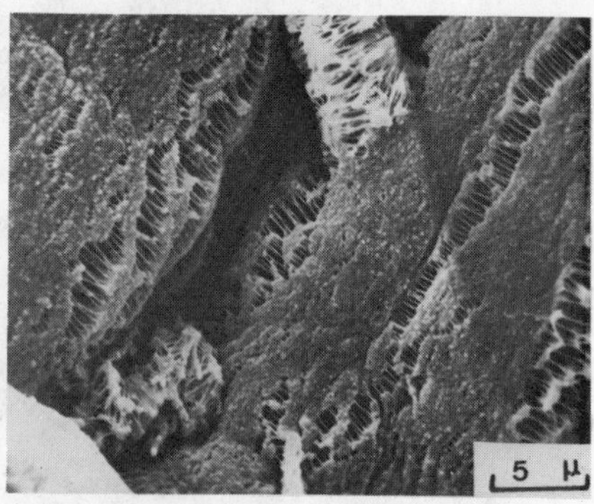

Figure 5. Scanning electron micrograph of the same sample as in Figure 4 but at higher magnification

TABLE I

WATER VAPOR TRANSMISSION OF ENCAPSULATED KRAFT PAPER

"Add On" %	Basis Weight g/M^2	Permeability (38^oC, 90% R.H.) $g/M^2/24h$.
0	67	4500
4	70	4300
9	73.5	4200
21	85	3900
29	89	3500

Details of these measurements have been reported elsewhere (5) but in summary one may say that all basic strength properties were preserved and not infrequently, significant improvements were noted. For example: breaking strength and elongation increased with add on as did "double fold" (M.I.T.). As was to be expected wet strength properties improved dramatically as can be seen in Fig. 6 for a sample with 20% polyethylene.

Once the scanning electron microscope had revealed the particulate nature of nascent polyethylene (Fig. 1c) one had to expect it to act as an organic pigment. TAPPI and printing opacity are significantly improved even for sheets which are initially of high brightness and this is attributable to the fine subdivision of the nascent polyethylene with particle sizes in the ideal size range for light scattering. When encapsulation was applied to a non-bleached paper pulp the whitening effects were rather spectacular as can be judged by Fig. 7.

CONTINUOUS ENCAPSULATION PROCESS FOR FIBROUS WEBS

Paper and textile treatments are generally continuous while polymerization is usually a batch process. In addition Ziegler-Natta catalyst is sensitive both to oxygen and water hence it seemed a rather formidable challenge to devevelop an on-line polyethylene treatment for paper and other web structures. Nevertheless the unique property improvements which were found in our preliminary batch studies on filter paper encouraged us to move into an area "where angels might fear to tread". If successful, a

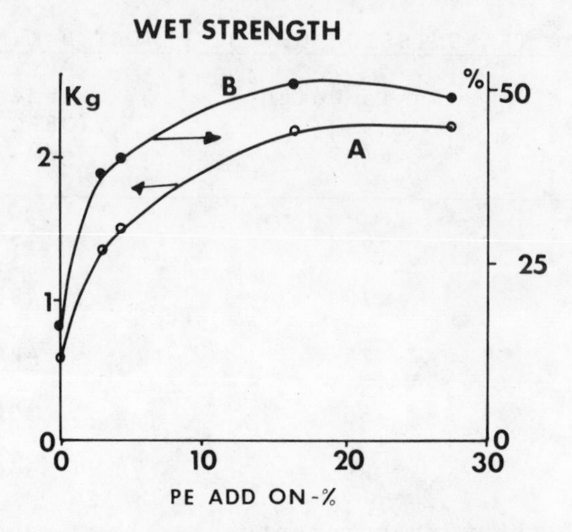

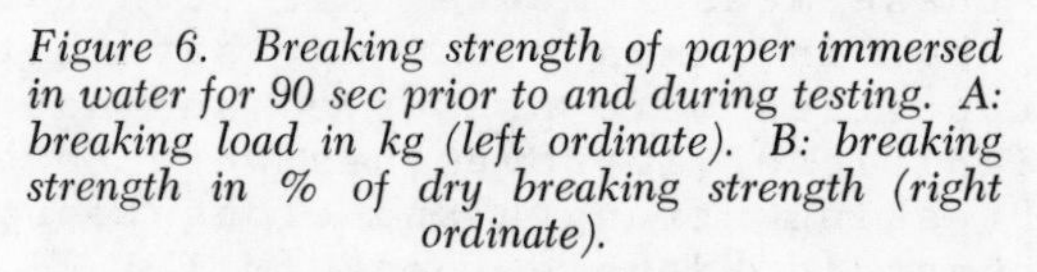
Figure 6. Breaking strength of paper immersed in water for 90 sec prior to and during testing. A: breaking load in kg (left ordinate). B: breaking strength in % of dry breaking strength (right ordinate).

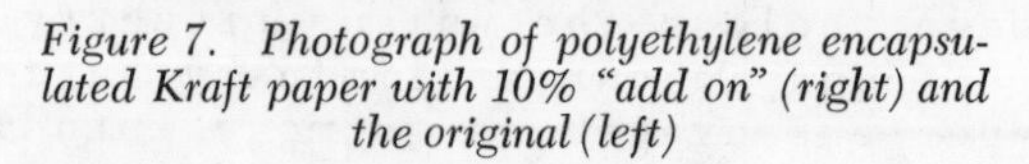
Figure 7. Photograph of polyethylene encapsulated Kraft paper with 10% "add on" (right) and the original (left)

continuous process would mean simultaneous polymerization and fabrication of the polyethylene into a useful continuous composite, something which is not realised to date in the field of transition metal catalysed polymerization of olefins where molding and extrusion follow polymerization as separate operations

The first requirement which was put on the process was that catalyst deposition should be rapid and uniform throughout the fibrous construct. The polymerization step was considered to be straightforward and sufficiently rapid. Our first prototype was operated inside a dry box to avoid O_2 and H_2O contamination of the catalyst. The second version which is shown schematically in Fig. 8 and photographically in Fig. 9 was a closed system which was extensively purged prior to admitting the chemicals for catalyst deposition and polymerization. The operating principles are standard in terms of wet treatments of papers and textiles i.e. a moving sheet dips successively into a series of baths. The key operation is the first one wherein the sheet moves into a 10^{-2}M. solution of $TiCl_4$ in isopentane and is then moved into a zone which allows preferential evaporation of the low boiling isopentane leaving behind finely distributed $TiCl_4$ in the fiber matrix. A good quality "bond" paper was used in all our early work and it was found that a speed of 20cm/min was the ideal machine speed for good catalyst deposition. Because of our present machine design, this became the limiting speed for the overall operation however it is a simple matter to avoid the speed controlling aspect of any given step in a continuous operation hence the focus for judging this work should be the continuous operation and not the speed.

Refering to Fig. 9, the steps from the first to the last reactor are the following:

- $TiCl_4$-isopentane sheet saturation (bath 1)
- isopentane evaporation
- $TiCl_3$ generation by action of $AlEt_3$
- polymerization

The success of the process is critically dependant on the care taken to avoid contamination of the different baths. An important step in this respect, besides the initial purge to eliminate O_2, was the addition of a lid directly over the reaction baths. This creates a hydrocarbon atmosphere in the critical zone and allows the polymerization reaction to proceed without contamination from catalyst poisons such as O_2 and H_2O which are brought into the apparatus by the paper roll. The latter is simply conditioned at room

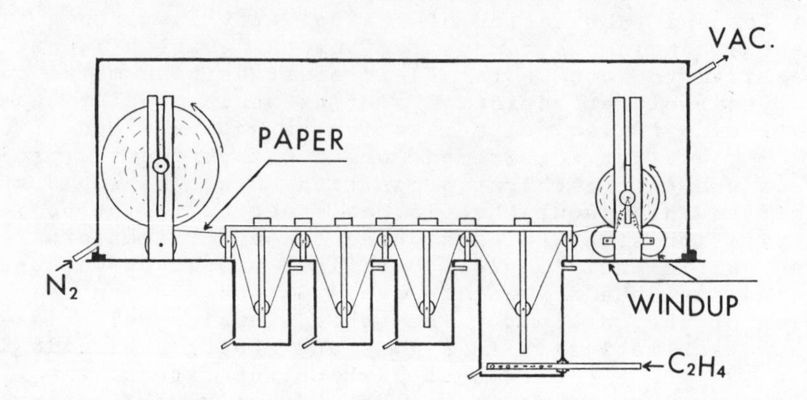

Figure 8. Schematic of the apparatus used for continuous encapsulation of paper with polyethylene

Figure 9. Photograph of the apparatus used for continuous encapsulation of paper with polyethylene

humidity and adsorbed H_2O is eliminated in the first bath by reaction with $TiCl_4$. This is not an economical nor ideal way of drying the substrate rather a rapid heating step prior to the first bath is recommended. It follows from our experience with cellulosics which are probably the most hydrophilic fibrous webs available that the process should be even better suited to non-moisture sorbing systems such as glass fibers.

Using the machine shown in Fig. 9, we have been able to treat a variety of paper substrates both bleached and unbleached. Simple adjustments in the operating variables allows us to control the locus of polymerization e.g. nascent polyethylene can be limited to the surface, thereby imitating a classical coating operation. On the other hand when polymerization is uniform throughout the matrix a unique composite is obtained: a bonded cellulose fiber web in a polymer continuous matrix. The latter preferentially surrounds the areas where the fibers are not bonded to each other.

Exploratory experiments have shown that encapsulated paper can be heat and pressure treated to yield new surface properties, heat sealability and changed mechanical properties. With respect to the latter it is important to realize that cellulose-cellulose bonds are maintained in this process as shown by the fact that the stress-strain curve of the encapsulate is almost identical to that of the original paper. Fig. 10 is a scanning electron micrograph of an encapsulated bond paper (20% polyethylene add on) which has been made glossy by heat-treating at $40 kg/cm^2$ and 100^oC; although the basic porosity of the encapsulating layer is still evident, it has decreased and the surface is much smoother than that of the original sheet.

CONCLUSIONS

The need for streamlining the numerous operations involved in various polymer processes is a constant goad to inventiveness for polymer scientists. The present development is no exception, although one aspect of this work should be underlined as being beyond this usual motivation: the nascent morphology of polymers, seldom turned to advantage by polymer chemists, is the essence of our process. The nascent morphology principle is used not only for localizing the polymer to the site where the catalyst was adsorbed but to provide some unique properties in the polymer matrix. The opacifying effect of the "pearl necklace" texture (cf. Fig. 1c) and the remarkable continuity of the polymer matrix which results from

Figure 10. Scanning electron micrograph of polyethylene encapsulated paper surface with 20% "add on" which has been heat-treated at 40kg/cm² and 100°C for 30 sec

the proliferous growth in the interfiber space are not usually used properties of synthetic polymers.

Like all processes which apply to a class of materials viz. fibrous aggregates, the operating conditions have to be optimized for each substrate. However the range of substrates studied so far is such as to allow the following conclusions:

- a minimum porosity level is required, textiles are ideal but certain types of papers can be too dense,
- predrying of the substrate is desirable,
- non-encapsulating polymer is not a problem unless some catalyst escapes from the sheet to become active hence the adsorbed catalyst should be well anchored,
- Ziegler-Natta catalyst can be handled in a continuous process such as this as long as O_2 contamination is avoided,
- on line texturizing is feasable,
- catalyst efficiency needs to be optimized in the present process preferably by devising a new method of deposition.

LITERATURE CITED

1. Herman, D.F., Kruse, V. and Brancato, J.J., J. Polym. Sci., C, 11, 75 (1965).
2. Dankovics, A., Erdelyi, J. and Koltai, M., J. Appl. Polym. Sci., 13, 1809 (1969).
3. Hitachi Ltd, Japan Patent 69.19.399 (1969).
4. Revol, J.F., Chanzy, H.D., Quere, J. and Guiroy, G., Fourth Canadian Wood Chemistry Symposium, Quebec, Canada, July 1973.
5. Fisa, B. and Marchessault, R.H., J. Appl. Polym. Sci., 18, 2025 (1974).
6. Kern, R.J. and Hurst, H.G., J. Polym. Sci., 44, 272 (1960).
7. Marchessault, R.H., Fisa, B. and Chanzy, H.D., Crit. Rev. Macromol. Chem., 1, 315 (1972).
8. Chanzy, H.D., Revol, J.F., Marchessault, R.H. and Lamandé, A., Kolloid Z. u. Z. Polymere, 251, 563 (1973).
9. Fisa, F., Revol, J.F. and Marchessault, R.H., J. Polym. Sci., Polym. Phys. Ed. (in press).

11

Formation and Properties of Blended Nonwovens Produced by Cellulose–Cellulose Bonding

RAYMOND A. YOUNG and BERNARD MILLER

Textile Research Institute, Princeton, N.J. 08540

Abstract

A series of self-bonded nonwoven fabrics have been produced from cellulosic fibers and from cellulose fiber/noncellulose (natural or manmade) fiber blends by treatment of cross-laid card webs with concentrated zinc chloride solutions. The cellulosic fibers become cohesively bonded during the treatment. If noncellulosic fibers have been blended in during the carding operation, their mechanical entanglement will be enhanced by this bonding. A number of parameters, such as zinc chloride concentration, can be varied to produce a range of desired physical properties and aesthetic characteristics. Usefulness of these nonwovens in filtration studies has been demonstrated.

Introduction

Nonwoven fabrics are being used on an increasing scale and in a widening variety of applications (1,2). Generally, the manufacture of these materials involves formation of a fibrous network or web, reinforcement of the web, and finishing of the product. The many methods for accomplishing this sequence may be classified on the basis of whether the web is wet-laid or dry-laid. In the wet process the fibers are suspended in water and processed essentially as in papermaking, whereas in the dry process the fibrous web is produced in the dry state either by carding or using special aerodynamic equipment designed to give an isotropic web.

One of three basic methods is usually employed to achieve bonding in dry-laid webs: mechanical entanglement, adhesive binding, or fiber fusion. To obtain strength through mechanical entanglement alone, needlepunching methods are most often used. Adhesive binding requires the addition of another substance to cement or bind fibers at crossover points. Polymeric binding agents used have included acrylics, polyvinyl chloride, polyvinyl acetate, cellulose esters, and polyurethanes. To a considerable extent, the characteristics of the final product depend on the fiber-binder ratio (2).

A growing number of fiber fusion techniques are becoming increasingly important in the production of nonwovens. Spunbonded fabrics, for example, are produced directly from extruded polymer stock. The extruded filaments are charged electrostatically so that they balloon, collected on a moving conveyer, and then thermally bonded (2,3). In other thermal fusion processes, fibers of a thermoplastic polymer are used as one component of the web, and bonding is achieved by heating or calendering at the appropriate temperature (4). Solvent fusion requires heating the web in the presence of a solvent that interacts with a web component (5).

Another method of fiber fusion is the autogeneous bonding of nylon fibers by treatment with gaseous hydrochloric acid (6). This treatment apparently causes surface decrystallization and plasticization of the nylon fibers and subsequent bonding upon removal of the acidic gas. The technique, however, is specific to nylon fibers and suffers the additional disadvantage that many other fibers cannot be used in blends with the nylon because of possible degradation during the gaseous HCl treatment.

The method described in this report for production of nonwoven fabrics can be classified as a fiber fusion technique. In this case, cellulosic fibers are partially decrystallized and solubilized by a concentrated aqueous solution of zinc chloride and are bonded on removal of the salt solution. In this way it is possible to bond cotton or rayon webs without the use of adhesive binders which frequently modify physical properties and which may alter significantly the chemical characteristics of the material (e.g., its flammability behavior). In addition, this method avoids the detrimental effects of high temperatures or chemical reagents which are normally necessary for fusion bonding. Consequently, this bonding method can be used both with totally cellulosic fiber systems and with most cellulosic/noncellulosic (natural or manmade) fiber blends.

Zinc chloride treatments have long been used for

bonding multiplies of cellulose sheets, resulting in rigid industrial materials whose physical properties resemble plastic laminates made from resin-paper systems (7,8). The use of concentrated aqueous zinc chloride solutions as an aftertreatment for enhancement of strength properties of previously bonded rayon nonwoven fabrics has also been described (9). Figure 1, taken from the work of Patil et al. (10), depicts the effect of zinc chloride solutions on the crystallinity of cotton cellulose as measured by x-ray fiber diffraction. Depending on the conditions of treatment, zinc chloride solutions can cause almost total decrystallization of a cellulose sample. The strength of a nonwoven web of cellulosic fibers treated with a concentrated solution of zinc chloride and then washed would presumably result from interfiber bonding accompanying recrystallization of the solubilized cellulose upon removal of the zinc chloride. This report summarizes recent studies on the use of such treatment to produce nonwoven fabrics from cross-laid webs of cellulose and cellulose/manmade fiber blends.

Experimental

Materials and Methods.

Manmade fibers. All the manmade fibers used in this study were crimped and had the following specifications: rayon - 5.5 denier, 1-1/16-in. staple length (FMC Corp.); polyester - 4.5 denier, 1-9/16-in. staple length (E. I. du Pont de Nemours & Co., Inc.); nylon - 3 denier, 1-1/2-in. staple length (E. I. du Pont de Nemours & Co., Inc.); acrylic - 2 denier, 2-in. staple length (Dow Badische Co.).

Cotton. Acala cotton card sliver was scoured with boiling 5% sodium hydroxide under reflux for 4 hr, washed thoroughly with water, and air dried. The air-dried fibers were formed into card webs without further treatment.

Zinc chloride. Standard solutions of zinc chloride were prepared by using the Fisher technical grade chemical. The concentration of the solutions was determined through measurements of specific gravities by weighing 25-ml of the solution to the nearest 0.1 mg in a sealed, 25-ml volumetric flask. The concentration of the solution was then obtained from a table of specific gravity versus concentration of zinc chloride as a function of temperature (11).

Formation of Nonwoven Mats. The fibers were first formed into a 15-in. wide card web and cut into four equal sections. (If more than one type of fiber was to be included in the mat, the blending of the fibers was done during the carding operation.) The four sections were cross-laid alternating the carding direction in each layer. Six-inch squares of this mat were saturated with concentrated aqueous zinc chloride, passed through a pressure nip to remove excess solution, and placed in a Carver press between Teflon® sheets. The mats were then pressed at 2 psi for one min at 100°C. Immediately after removal from the press, the zinc chloride was leached from the fabric in a running water bath. The excess water was then pressed out and the mat dried in the Carver press at 100°C at 6 psi. The final thickness and density could be adjusted to some extent during the final pressing.

Results and Discussion

Strength Properties. The earlier work of Patil et al. (10) (Fig. 1) demonstrated the critical range of concentration necessary for decrystallization of cotton cellulose. This concentration dependence probably is the result of the marked effect of ionic distribution on the assumed zinc chloride - cellulose complex (12). The zinc chloride solution has essentially no effect on the cellulose crystal structure either below a concentration of 60% (W/W) or above 75%. The curves in Figure 1 also indicate that cotton decrystallization is extremely sensitive to relatively small changes in temperature.

Figure 2 depicts tensile strength as a function of zinc chloride concentration for a series of carded cotton webs bonded according to the procedure described in the previous section. The effect of zinc chloride concentration on the tensile strength of the cotton nonwovens closely parallels the reported effect on the extent of decrystallization of cotton (Fig. 1). The maximum strength was achieved at about 72-73% zinc chloride (Fig. 2), the same range observed for maximum decrystallization.

Figure 3 shows the tensile strengths of a 100% rayon sample and of a rayon/polyester (75/25) blend as a function of zinc chloride concentration. The maximum strengths of both rayon-based nonwovens are

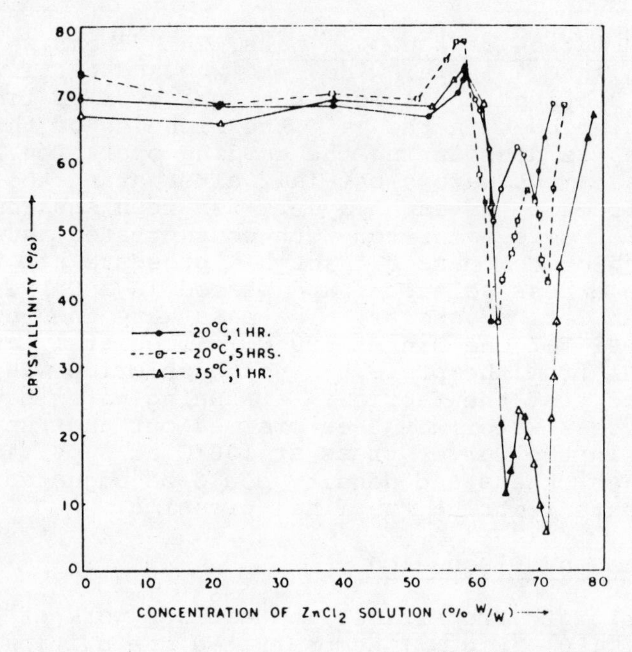

Textile Research Journal

Figure 1. Decrystallization of cotton by zinc chloride (10)

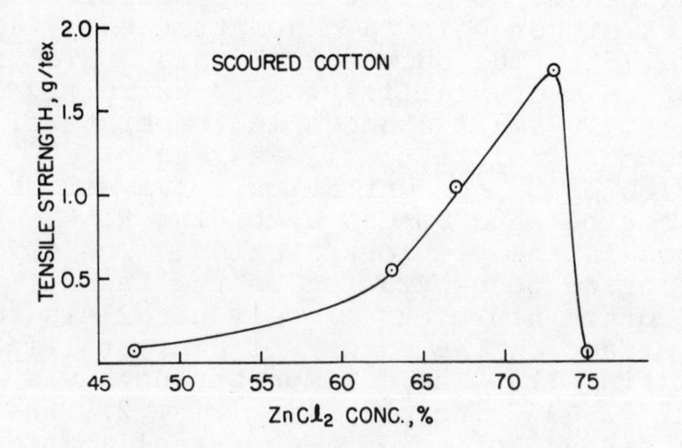

Figure 2. Tensile strength of zinc chloride-bonded scoured cotton nonwovens as a function of zinc chloride concentration

about half that of the cotton nonwoven. These strength differences probably reflect differences in fiber strength, the cotton having 1.5 to 2 times the tenacity of the rayon fiber (13). In addition, it appears that a much lower concentration of zinc chloride, about 56%, is necessary for obtaining maximum strength with rayon-based systems. This difference is most likely a reflection of differences in the crystalline character of the two types of cellulose fibers. Rayon has a crystallinity of about 40% and cotton close to 70% as determined by x-ray fiber diffraction (14). Furthermore, rayon has a cellulose II crystalline lattice structure, while cotton has the native cellulose I crystalline lattice structure.

At very high concentrations of zinc chloride, the rayon was found to form a weak gel-like film under the bonding conditions used in this study (Fig. 3). Film formation under these conditions is the result of total decrystallization and loss of rayon fiber structure. This illustrates the principle that in order to achieve textile-like nonwovens of acceptable strength, bulk decrystallization is to be avoided and fiber structure must be maintained. Thus, the zinc chloride treatment must induce only surface decrystallization of the cellulosic fibers. The strength of the nonwoven can then be maximized through an optimum combination of extent of bonding and mechanical entanglement. With the blended systems, the strength of the nonwoven mat results from both bonding and mechanical entanglement of the cellulosic fiber, and from purely mechanical entanglement of the noncellulosic fiber component. The higher strength of the rayon/polyester blend (Fig. 3) reflects the higher strength of the mechanically entangled polyester component.

The tensile strengths of two other blended systems, rayon/nylon and rayon/acrylic, as a function of the rayon content in the blend are shown in Figure 4. It was expected that as the amount of rayon in the blend increased, the tensile strength would also progressively increase. This tendency was observed for the rayon/nylon blend (Fig. 4), where the strength of the blend increased for a blend containing 60% rayon to 0.9 g/tex, the same strength measured for the 100% rayon system. It is likely that a maximum tensile strength for rayon/nylon occurs somewhere between 60 and 100% rayon. This would be consistent with the results shown in Figure 3 for rayon/polyester. The

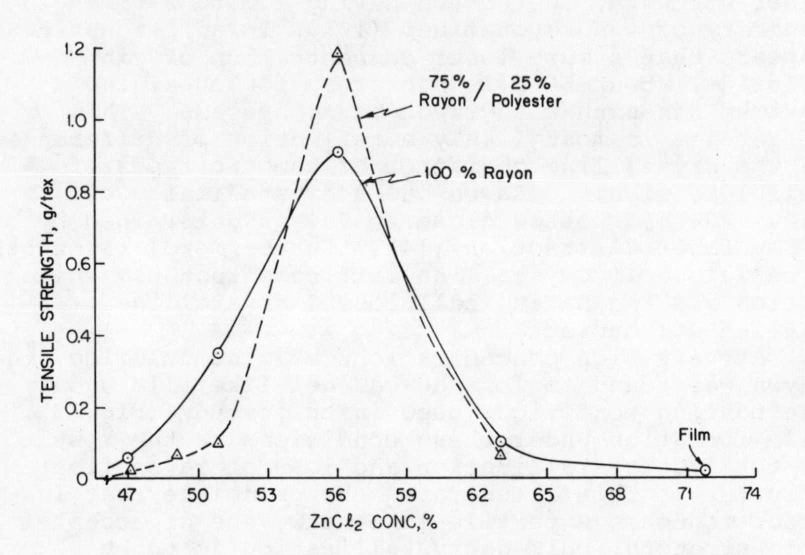

Figure 3. Tensile strength of zinc chloride-bonded nonwovens as a function of zinc chloride concentration

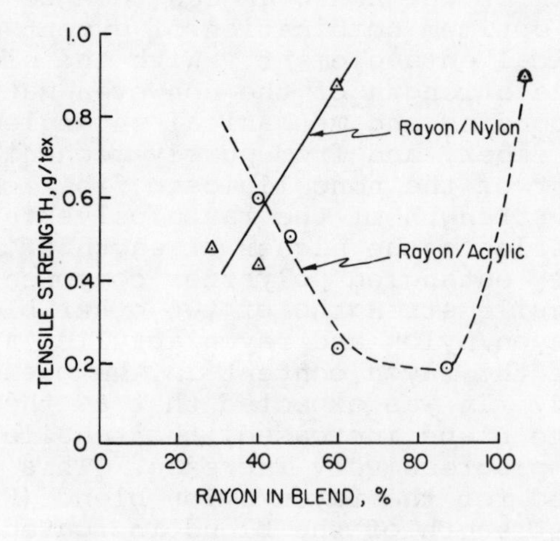

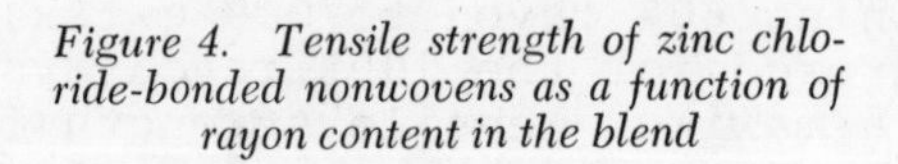

Figure 4. Tensile strength of zinc chloride-bonded nonwovens as a function of rayon content in the blend

rayon/acrylic blend, however, behaved in a different manner. Instead of showing enhanced tensile strength with increasing amounts of rayon in the blend, this system showed an opposite correlation. This was most likely caused by interaction of the acrylic fiber with the zinc chloride solution, since these solutions are at least partial solvents for polyacrylonitrile polymers. Under the bonding conditions used in these experiments, it was found that with a 100% acrylic mat a filmy material was formed after zinc chloride treatments. This acrylic film in combination with small amounts of rayon formed a strong uniform mat. However, at lower levels of acrylic fiber in the blend, the acrylic film exhibited nonuniform shrinkage creating gaps in the nonwoven structure which reduced the strength properties. The possible utilization of the interaction of acrylic fibers with concentrated aqueous solutions of zinc chloride for bonding of nonwoven webs is under further investigation.

Electron Microscopy. Figures 5 and 6 show scanning electron micrographs of sections of a rayon nonwoven mat bonded with 56% zinc chloride under the standard conditions. Several fiber crossings are shown in Figure 5A (300X) with one of these exhibiting apparent bonding between the fibers. The occasional bonding which exists in these structures is mainly responsible for the flexibility needed in the nonwoven system, since excessive bonding gives a stiff board-like material. Under higher magnification (1000X) the nature of the bonded area is clarified (Fig. 5B). Though the fiber structure appears intact, there has been extensive surface decrystallization and solubilization causing a melding of the fibers by a cellulose-cellulose bond, which forms after leaching removes the zinc chloride.

Figure 6 shows electron micrographs of a delaminated zinc chloride-bonded rayon nonwoven. Figure 6A (300X) depicts a location where apparently there has been fiber-fiber pullout at a bonded point as a result of the delamination. Also shown are fibers where no bonds were formed but which add strength through mechanical entanglement. At higher magnification (1000X) the filmy character of the broken bond becomes evident. The decrystallization and solubilization is apparently localized to some extent at fiber-fiber crossings, since the remainder of the fiber shows only lesser effects of the zinc chloride treatment (also cf. Fig. 5A).

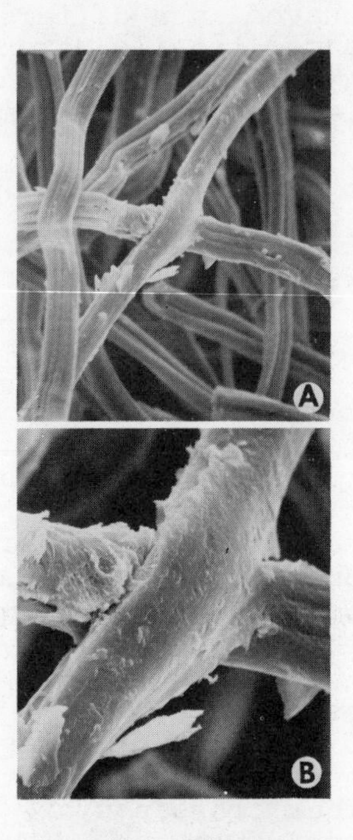

Figure 5. Scanning electron micrographs of sections of a zinc chloride-bonded rayon nonwoven (A—128×, B—425×)

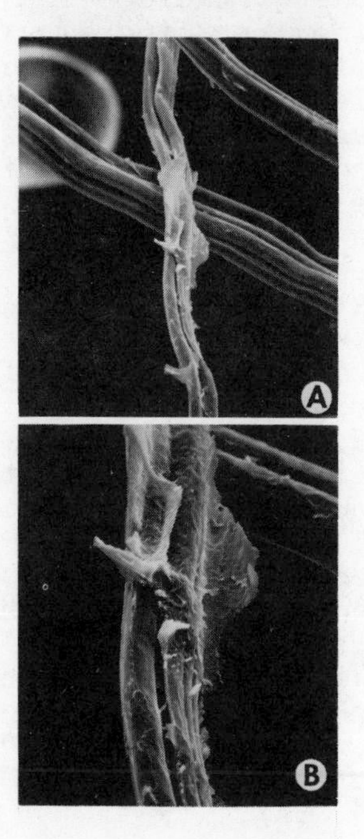

Figure 6. Scanning electron micrographs of a delaminated zinc chloride-bonded rayon nonwoven (A—128×, B—425×)

Air Particulate Filtration. One of the many applications for nonwoven fabrics is their use as air filters. A series of preliminary experiments using the cellulose-based, zinc chloride-bonded fabrics as filter media have been conducted in the filtration laboratory at TRI (15). Figure 7 shows filtration efficiencies as a function of mat density for several of the zinc chloride-bonded fabrics. At very high densities all the fabrics give essentially the same high efficiency; however, this is trivial, since the excessive drag and the resultant energy consumption would be unacceptable. In terms of filtration performance, it is necessary to develop filters that give high efficiency with minimal drag (i.e., low density).

Distinct differences are noted among the different fiber types with fabrics of lower densities. Cotton maintains the highest efficiency at all density levels, which is probably due to the much lower linear density of the cotton fiber as compared to that of the rayon (5.5 denier) and polyester (4.5 denier) fibers. Thus, with the cotton mat at a given density or weight there are many more fibers resulting in many more fiber crossings and channels per unit area to give greater efficiency. On the other hand, the rayon and polyester fibers have about the same dimensions. The higher efficiency of the rayon/polyester blend as compared to that of the all rayon must, therefore, be in part related to the different surface characteristics of the polyester component which apparently enhances filtration performance. Studies of the use of other blended fiber systems for filtration applications are in progress.

Literature Cited

1. Carter, H. J., Tappi, (1974), 57, 50.
2. Ross, S. E., Chem. Tech., (1972), 2, 535.
3. Ericson, C. E., "Bonding in Spunbonded Nonwovens", Textile Research Institute Seminar, October 21, 1974.
4. Winchester, S. C. and Whitwell, J. C., Textile Res. J., (1970), 40, 458-471.
5. Technical Bulletin, "Production of Nonwoven Fabrics", Badische Anilin and Soda-Fabrik AG, (BASF).
6. Mallonee, W. C. and Harris, H. E., Gas Activated Bonding of Nylons, U.S. Patent 3,516,900 (June 23, 1970).

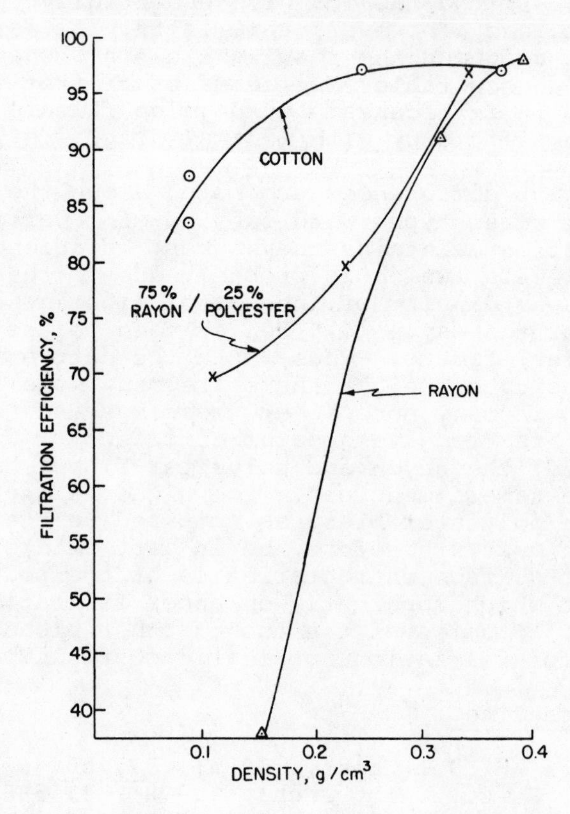

Figure 7. Filtration efficiency of zinc chloride-bonded nonwovens as a function of mat density

7. Taylor, T., Brit. Pat. 787 (March 29, 1859); U.S. Pat. 114, 830 (March 16, 1871).
8. Lee, L. T. C., Tappi, (1964), 47, 386.
9. Tech. Ser. Bull., S-12R, American Viscose Div., FMC Corp., Marcus Hook, Pa., July 15, 1968.
10. Patil, N. B., Dweltz, N. E., and Radhakrishnan, T., Textile Res. J., (1965), 35, 517.
11. International Critical Tables, 1st ed., Vol. III, McGraw-Hill, New York, 1928, p. 64.
12. Richard, N. J. and Williams, D. G., Carbohyd. Res. (1970), 12, 409.
13. Morton, W. E. and Hearle, J. W. S., "Physical Properties of Textile Fibers", p. 285, Butterworth Ltd., London, 1962.
14. Ott, E., Spurlin, A. M., and Grafflin, M. W., eds., Cellulose and Cellulose Derivatives, Part I, New York, Interscience Pub., 1954, p. 273.
15. Miller, B., Lamb, G. E. R., and Costanza, P., "Nonwovens, A Study of Critical Fiber Variables", Tappi, in press.

12

Marine Polymers, V. Modification of Paper with Partially Deacetylated Chitin

G. G. ALLAN, J. F. FRIEDHOFF, M. KORPELA, J. E. LAINE, and J. C. POWELL

College of Forest Resources, University of Washington, Seattle, Wash. 98195

The low level incorporation of partially deacetylated crab chitin or crabshells increases the printing opacity of paper. The tensile strength of paper from unbeaten, but not beaten, pulp is concurrently improved. The magnitude of the strength improvement is related inversely to the size of the particles and directly to the extent of deacetylation. Handsheets from beaten stock are simultaneously strengthened and opacified by the inclusion of papermaking grade Kaolin clay coated with chitosan.

Crab and shrimp shell wastes are an abundant source (1) of chitin (Fig. 1, a) which when deacetylated affords the polyamine, chitosan (b). Although this macromolecule is an outstanding binder for cellulosic (c) fiber structures (2), its isolation from crustacean residues is somewhat tedious and the deacetylation step results in a weight loss of slightly more than 20%. Moreover, during this process the intimately admixed calcium carbonate content of the laboriously harvested shells is usually converted into a low value soluble calcium salt with the concomitant consumption of acid and a further weight loss. Acidolysis also induces a degradation in the molecular weight of the chitosan which adversely affects its performance as a paper additive (3). For these reasons, attention is now being directed towards an evaluation of the utility of chitin itself as well as

---

Address inquiries to Prof. G.G. Allan, AR-10, University of Washington, Seattle, WA 98195.

the chitin-protein-calcium carbonate complex of the shell as it exists in the shell before decalcification. This article reports such an assessment in the paper additive area.

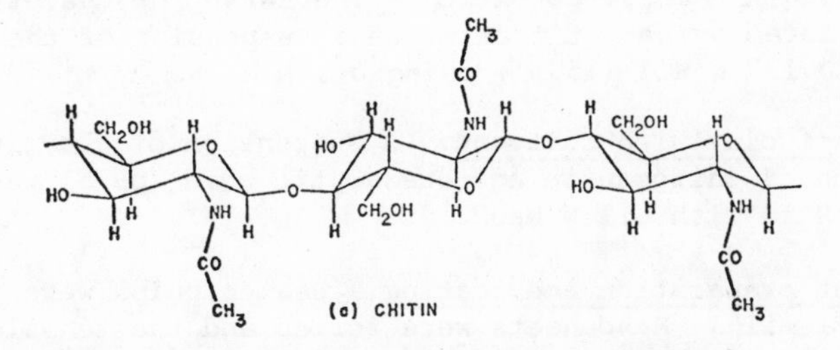

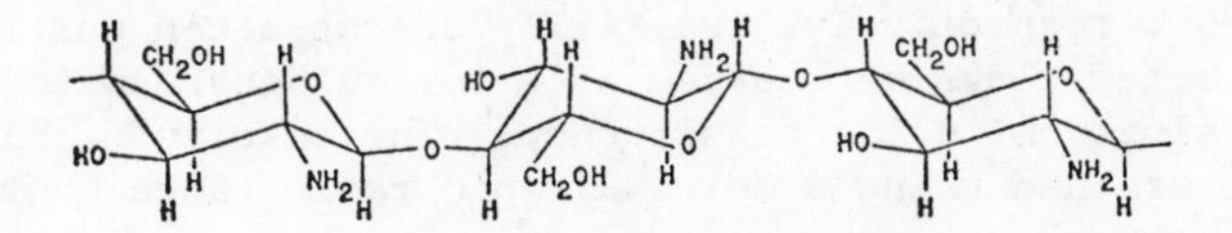

(b) CHITOSAN

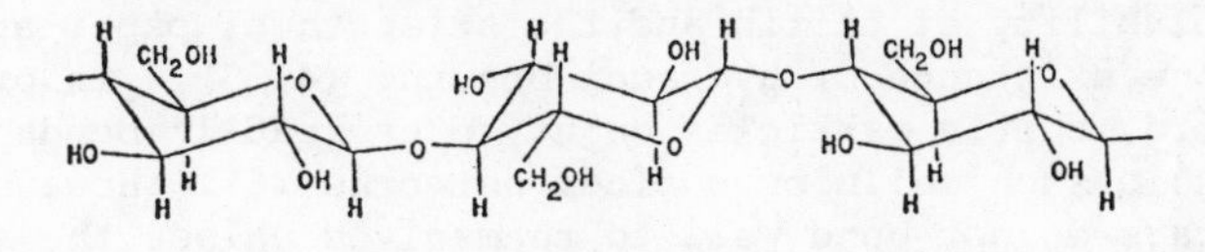

(c) CELLULOSE

Figure 1. Segments of chitin, chitosan, and cellulose

Experimental

Materials. The materials used were western hemlock (*Tsuga canadensis*) sulfite pulp provided by the Weyerhaeuser Company, Seattle; crabshell, chitin and chitosan supplied by the Food, Chemical and Research Laboratories, Inc., Seattle; and Astra-glaze, a clay furnished by the Georgia Kaolin Company, Elizabeth, New Jersey. All other chemicals were standard stock items.

Methods.

Grinding and size classification of chitin and crabshell particles. Both the chitin and crabshells were supplied as small flakes which were subsequently ball-milled for 3 days with glass marbles. The product was classified using Tyler Standard screens (140, 200 and 325 mesh) to yield fractions 74-105μ, 44-74μ and <44μ in size.

Deacetylation of chitin and crabshell particles. Chitin or crabshell particles (5g) were refluxed with a 30% aqueous sodium hydroxide solution (100ml) for 15, 30, 60, 120 or 240 min, filtered and washed with distilled water (700ml). The product was stored and used in a moist condition. The extent of deacetylation was estimated by back titration of a suspension of the product (1g) in 0.137 N HCl (150ml) using 0.1 N NaOH.

Coating of clay with chitosan. A suspension of clay in a 1 or 2% solution of chitosan in aqueous acetic acid (pH 5) was adjusted to pH 10 with 0.1 N NaOH.

Handsheet preparation and testing. Beaten pulps were prepared in a PFI mill. Handsheets were formed and the tensile properties measured in accordance with Tappi Standards T-205 m-58 and T-220 os-60, respectively. Partially deacetylated particles or chitosan-coated clays were added as water suspensions to a stirred pulp slurry of 0.15% consistency. The opacity of handsheets was determined using a Zeiss Elrepho reflectance photometer.

Results and Discussion

The insolubility of chitin and the strength of paper are both due to the existence of hydrogen bonding (4, 5). In principle therefore, chitin particles ought to be readily bondable to fibers within such cellulosic fiber networks (6). However, even pulp fibers do not bond well to themselves unless the area of contact has been increased by the mechanical process of beating which flexibilizes the hollow tubular structure (7) and fibrillates the exterior surfaces (8) of the fibers. In the case of chitin particles the fistular structure is not present and improved contact must therefore be sought by fibrillation. Since the beating process is notoriously inefficient (9) modification of the chitinous surfaces was instead explored by means of alkaline hydrolysis. Presumably this would result in some deacetylation which would have the effect of creating chitosan-type chains on the exterior of the chitin particles. These more mobile macromolecular appendages could presumably then more readily participate in fiber-particle bonding in a fashion analogous to the situation in water clarification where insoluble suspended particles are bridged and collected by water soluble polymers (10). This anticipated behavior was indeed observed when crab chitin, deacetylated to varying degrees, was incorporated into handsheets. The data obtained and summarized in Figs. 2 and 3 show clearly that as free amino groups are generated by the hydrolysis of the surface of the chitin particles, the tensile strength of the corresponding handsheets is increased. This presumably reflects the improved bonding capacity of the chitosan-like moieties which have an ionic bonding capability absent in chitin (11).

Nonetheless, in spite of any augmented reactivity of the chitin particle surface its overall dimensions will be relatively unchanged and the size of a paper additive will have a profound effect on its behavior within the sheet assuming that the amount present is less than the normal void volume within the sheet (12). Thus, if the particle diameter is larger than the thickness of the fiber then the number of fiber crossings will probably be reduced and fiber-fiber bonding will be hindered.

On the other hand, if the particle diameter is smaller than the thickness of the fiber then fiber-fiber interactions should not be so adversely effected because the particles will be predominantly positioned within the interstices of the sheet. In these locations the particles could even make some contribution to the strength of the sheet. Of course once the quantity of the additive approaches or exceeds the void capacity of the paper then fiber-fiber interactions will probably be inhibited by all sizes of particles.

These views are supported by the tensile strength data for handsheets containing partially deacetylated chitin particles of various sizes which is collected in Table 1. It is apparent that the addition of the reactive particles produced no major increase in the breaking length of the sheet until the diameter of the particles was less than 44μ. It is significant that the hemlock fibers used have an average diameter of 45μ (13). Even so, the initial beneficial effect of these smallest particles is ultimately offset by the interference with fiber-fiber bonding which, from Fig. 3, begins when the level of addition exceeds 3%. If the handsheet stock is beaten, the admixture of deacetylated chitin particles did not strengthen the resultant paper and the diminution in breaking length is shown in Fig. 4. This adverse effect is probably related to the dimensions of the deacetylated chitin particles which are large in terms of the individual voids within the structure of the handsheet made from the more conformable beaten fibers. Evidence supportive of this interpretation was adduced by a study of a papermaking grade clay which had been coated with chitosan by a precipitation technique (3). Such clays consist of particles much finer than the chitin derived material. A typical size range would be 2-5μ (14). Thus, the formation and tensile testing of handsheets from a mixture of pulp, beaten for 1000 revs in the PFI mill, and the modified clay gave the results presented in Fig. 5. This data shows that significant increases in breaking lengths are obtained at both the 10 and 20% level of filler addition when the clay particles are surrounded by precipitated chitosan.

Table 1. Properties of handsheets made from unbeaten pulp containing surface deacetylated chitin or crabshell particles.

Additive				Tensile properties	
description	size range	deacetylation time	loading	breaking length	increase in breaking length
Partially deacetylated crab chitin	74-105μ	15 min	2%	1190 m	10.3%
	74-105	30	2	1150	6.5
	74-105	60	2	1110	2.8
	44-74	15	2	1240	14.8
	44-74	30	2	1210	12.0
	44-74	60	2	1190	10.3
	<44	15	3	1240	14.8
	<44	30	3	1320	22.2
	<44	60	3	1470	36.1
	<44	120	2	1240	14.8
	<44	240	3	1280	18.5
Partially deacetylated crabshell	<44	60	3	1140	5.6
None	-	-	-	1080	-

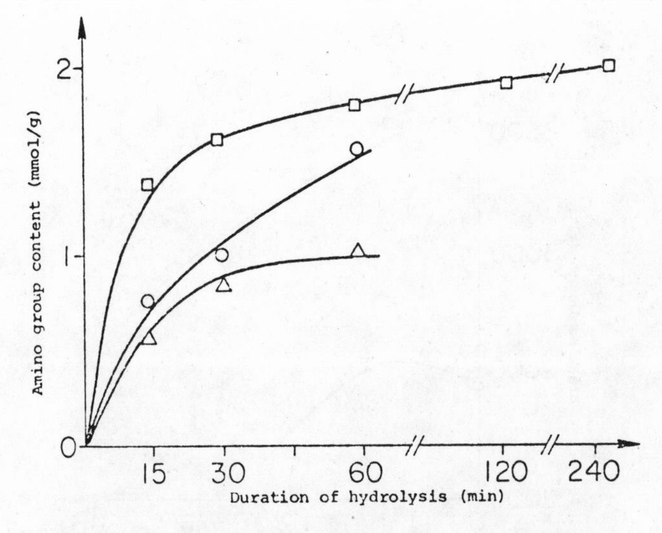

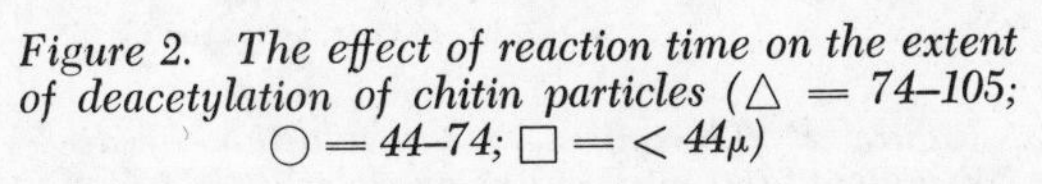
Figure 2. The effect of reaction time on the extent of deacetylation of chitin particles (△ = 74–105; ○ = 44–74; □ = < 44μ)

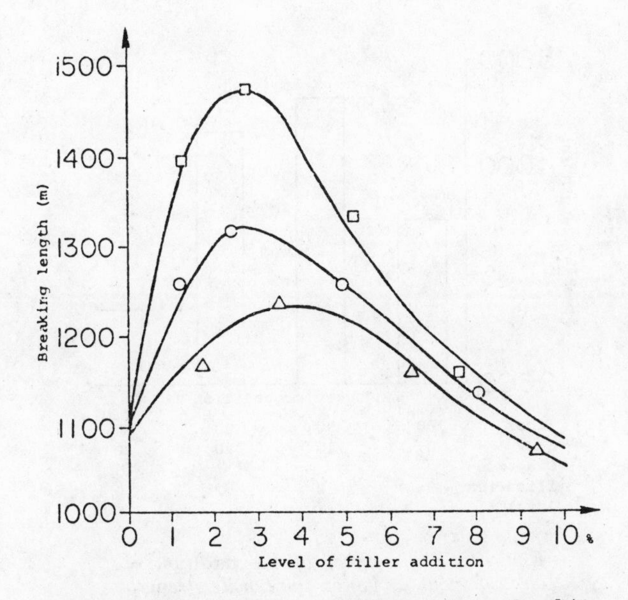

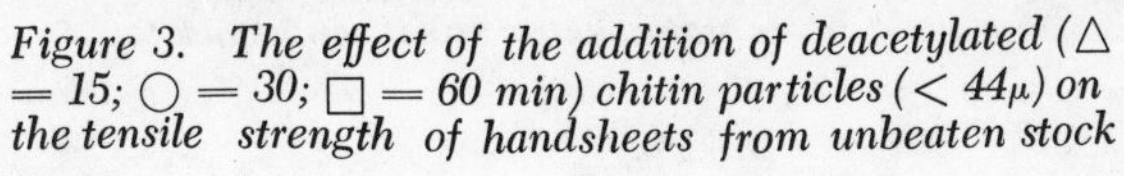
Figure 3. The effect of the addition of deacetylated (△ = 15; ○ = 30; □ = 60 min) chitin particles (< 44μ) on the tensile strength of handsheets from unbeaten stock

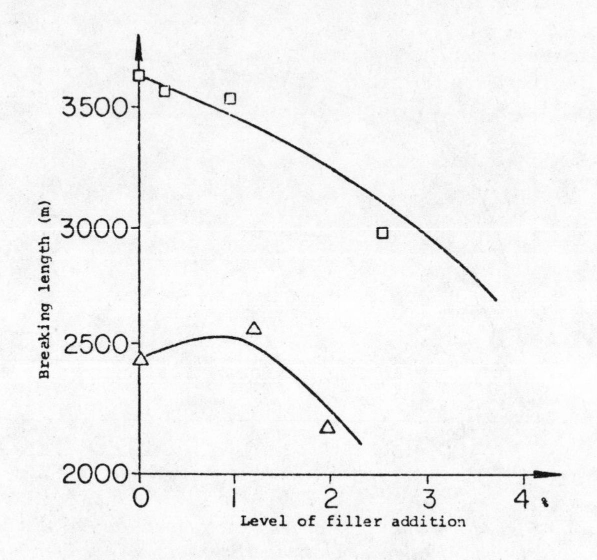

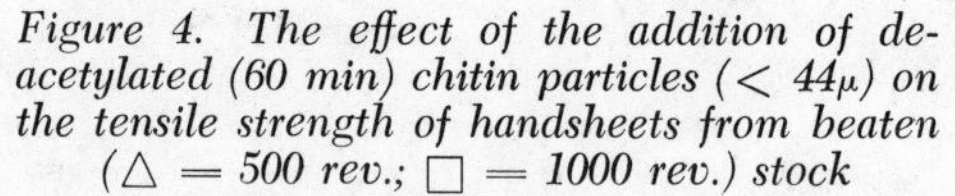
Figure 4. The effect of the addition of deacetylated (60 min) chitin particles (< 44μ) on the tensile strength of handsheets from beaten (△ = 500 rev.; □ = 1000 rev.) stock

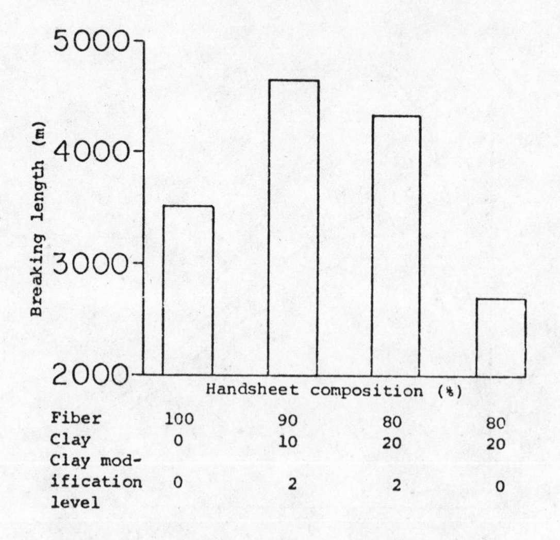

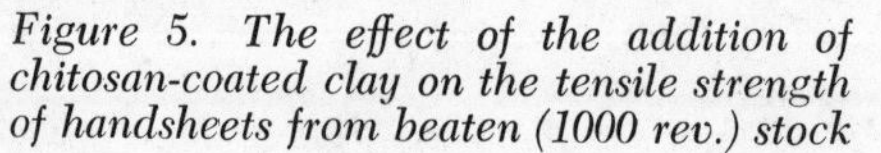
Figure 5. The effect of the addition of chitosan-coated clay on the tensile strength of handsheets from beaten (1000 rev.) stock

These strength improvements are accompanied by increases in the printing opacity (R_O/R_∞) of the sheets as exemplified by the data in Fig. 6. The polymer chitosan itself, of course, is known to improve the Z-strength, and concomitantly, the printability of paper (3). Surprisingly, as depicted in Fig. 7, both chitin and calciferous proteinaceous crabshells which have been ground to <325 mesh and partially deacetylated improve the printing opacity to about the same extent as the much finer kaolin clay but only up to the 2% level of addition. Thereafter further additions of the marine materials are less effective than the kaolin clay.

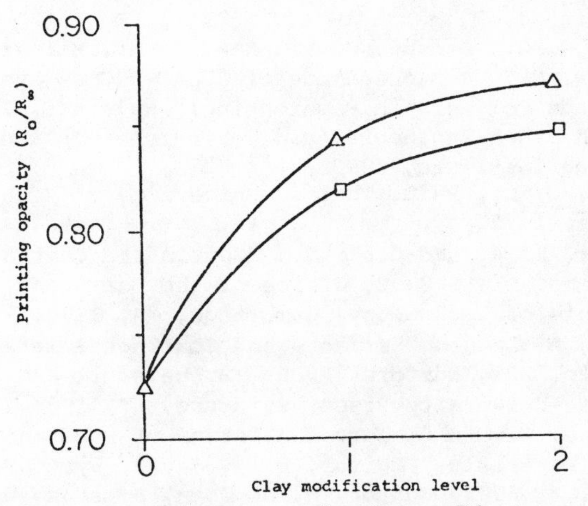

Figure 6. The effect of the clay modification level of the printing opacity of handsheets (clay content: □ = 10%; △ = 20%)

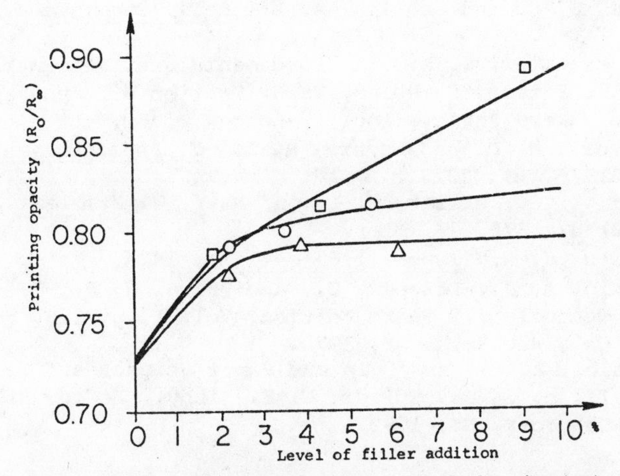

Figure 7. The effect of the addition of Kaolin clay (□), partially deacetylated crab chitin (○), or partially deacetylated crabshell (△) on the printing opacity of handsheets

Acknowledgment

The research reported in this publication was supported by the Washington Sea Grant Program, a part of the National Grant Program, which agency is maintained by the National Oceanic and Atmospheric Administration of the U.S. Department of Commerce.

Literature Cited

1 Mendenhall, V., "Utilization and Disposal of Crab and Shrimp Wastes," University of Alaska Sea Grant Marine Advisory Bulletin No. 2, March 1971.
2 Allan, G. G., Crosby, G. D., Lee, J.-h., Miller, M. L., and Reif, W. M., in Proceedings of IUPAC-EUCEPA Symposium on "Man-Made Polymers in Papermaking," Helsinki, 1972, p. 85. Finnish Paper Engineers Assoc.-Finnish Pulp and Paper Research Institute, 1973.
3 Crosby, G. D., Ph.D. Thesis, University of Washington, Seattle, 1973.
4 Pariser, E. R. and Dock, S., "Chitin and Chitin Derivatives," Report No. MITSG 73-2, Office of the Sea Grant, Massachusetts Institute of Technology, Cambridge, MA, 1972.
5 Nissan, A. H., in "Surfaces and Coatings Related to Paper and Wood," p. 221, Editors, R. H. Marchessault and C. Skaar, Syracuse University Press, Syracuse, NY, 1967.
6 Allan, G. G., in "Theory and Design of Wood and Fiber Composite Materials," Editor, B. A. Jayne, Syracuse Wood Science Series, 3, 299. Editor, W. A. Côté, Syracuse University Press, 1972.
7 Allan, G. G., Miller, M. L. and Neogi, A. N., Cell. Chem. Tech., (1970) 4, 567.
8 Stone, J. E. and Scallan, A. M., Cell. Chem. Tech., (1968) 2, 343.
9 Van der Akker, J. A., in "Fundamentals of Papermaking Fibers" p. 435, Editor, F. Bolam, Tech. Section British Paper and Board Makers Assoc. (Inc.). London, 1961.
10 Ries Jr., H. E. and Meyers, B. L., J. Appl. Polymer Sci. (1971) 15 (8), 2023.
11 Allan, G. G., Miller, M. L. and Reif, W. M., Text. Res. J., (1972) 42, 675.
12 Swanson, J. W., Tappi (1961) 44 (1), 142A.
13 Panshin, A. J., DeZeeuw, C., and Brown, H. P., "Textbook of Wood Technology," Third Edition, Vol. I, p. 123, McGraw-Hill Book Co., New York, NY, 1970.
14 Schwalbe, A. C., in "Pulp and Paper Science and Technology," Vol. II, p. 64-67, Editor, C. E. Libby. McGraw-Hill Book Co., New York, NY, 1962.

13

Application of GPC to Studies of the Viscose Process Part V. Effect of Heat on Rayon Properties

JOHN DYER

FMC Corp., Fiber Division, R&D, Marcus Hook, Penn. 19061

Summary

In this paper, results from studies of the effect of heating regenerated cellulose fibers having all-skin and all-core (not polynosic) structures on some fiber properties are discussed. The studies were made by heating the fibers at 184°C in the presence of air for 0, 1, 3 and 16 hours. The properties measured included wet and conditioned tenacity, single fiber flex life, water imbibition, moisture regain, basic degree of polymerization, cross-section staining characteristics and the molecular weight distribution by gel permeation chromatography. Results indicated that fiber properties were influenced by structure, basic DP and molecular weight distribution; i.e. by the size of cellulose molecules and the way they are arranged and held together in the fiber. Many of the property changes could be related to structural change and degradation that occurred on heating. For most of the heated samples, the weight average DP determined by GPC was significantly greater than the value determined viscosimetrically. Explanations are suggested for these observations.

Introduction

Rayon is a manufactured fiber composed of regenerated cellulose. Many types having different properties can be made from cellulose wood pulp by the viscose rayon process. Some examples of commercial fibers and their properties are shown in Table 1. The properties are determined by the size of the cellulose molecules and by the way they are arranged and held together in the fiber. Changes in the fiber structure caused by conditions of end-use alter the properties and

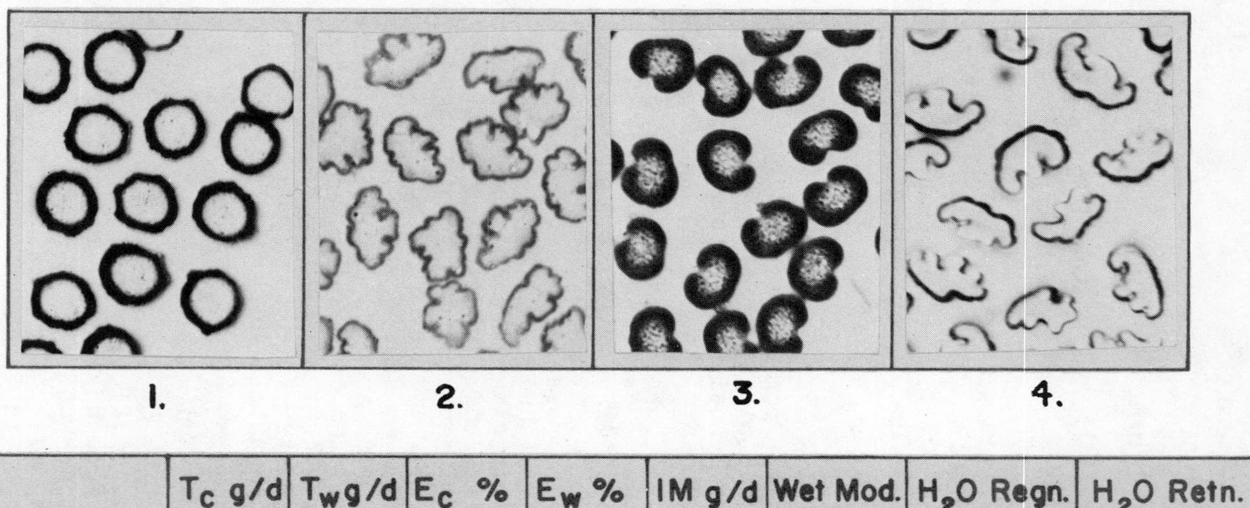

	T_c g/d	T_w g/d	E_c %	E_w %	IM g/d	Wet Mod.	H_2O Regn.	H_2O Retn.
1. H W M	3.5-6.0	2.5-4.6	8-18	15-22	85-110	9-25	10-12	60-80
2. REGULAR	2.2-3.5	0.5-2.2	10-30	22-35	60-90	3-4	11-13	90-115
3. TIRE	3.4-5.3	2.1-4.2	10-14	21-25	105-125		13-15	70-90
4. CRIMP	2.7	1.3	20	27	80	4	12	100

TABLE 1 SOME TYPICAL RAYON PROPERTIES

influence the useful life of the product.

The most usual properties determined from the response to tensile stress are the tenacity, extensibility, modulus and breaking energy - the work to rupture calculated from the area beneath the stress-strain curve. As might be expected, these can be related to structural parameters such as the degree of polymerization (DP), crystallinity, and molecular orientation. Additional information about structure can be obtained by differential staining of cross sections and from water imbibition and moisture regain measurements. But the correlation of fiber properties with structure is extremely complex. It is virtually impossible to change one variable without affecting others.

In many end-use applications, heat is produced by various mechanical processes such as bending, flexing, stressing and abrading, and by sunlight and other irradiation. Appreciable thermal exposure can also occur during the care cycle such as in laundering and drying. In industrial and commercial applications, e.g. conveyor belts, drive belts and tires, the fibers are often subjected to much higher temperature and without proper insulation and protection will degrade.

In studies of the effect of heat on rayon properties, it has been the usual practice to follow weight losses under different conditions with or without identification of the degradation products. This information has then been used with measurements on various physical and chemical properties of the fiber to explain the structural and chemical changes that occur on heating.

Without a tedious fractionation, only very limited information about the size of the cellulose molecules can be obtained from viscosity measurements.

Gel permeation chromatography has been used to measure the DP distribution of cellulose at various stages of the viscose rayon process and the changes that occur under conditions simulating end-use. (1,2,3) In this study of the effect of heat on rayon properties, evidence has been obtained that data from the GPC measurement can be influenced by differences in fiber structure.

The existence of two structurally different regions in the fiber cross section of Table 1 has been revealed by a differential staining technique (4). They are known as skin and core. Skin is normally found at and near the surface of the filament and surrounds the core. It is possible to vary the structure from all-core to all-skin, depending on the conditions used to make the fiber. The question arose whether skin and core would have different thermal response.

It was recognized that the response of the fiber to various treatments would also depend on other factors such as molecular orientation, molecular size, the total crystallinity and distribution of ordered and disordered regions. In this paper, only the all-core (excluding polynosic) and all-skin fibers will be compared with results being obtained for three samples of each type. It is emphasized that the experimental fibers, which are at the extremes of the structural limits, deviate in properties from commercial types.

Experimental

The experiments were made by heating the fiber at 184°C in an air oven. The sample as a continuous fiber bundle (214 filaments, 330 total denier) was held between two 7-inch stainless steel discs placed 12 inches apart by looping over a series of 240 notches cut in the edge of the discs. Before loading, 1300 gram compression was applied to the spring-loaded top disc which was then clamped in place. After loading, the top disc was released and the system allowed to equilibrate before placing in an air oven set at 184°C. There was some heat loss on placing the sample in the oven, the system taking about 20 minutes to regain thermal equilibrium. This temperature profile was similar for all the samples. After treatment, the sample was allowed to cool to room temperature. Samples for analysis and testing were cut as 10-inch lengths from between the discs, discarding the fiber that had been in contact with the metal.

For the molecular weight distribution measurement, the samples were nitrated and dissolved in tetrahydrofuran. Fractionation was achieved on the basis of size alone using a Waters 100 GPC unit with a series of four columns containing crosslinked polystyrene resin beads of 10^6, 10^5, 10^4 and 10^3A° pore size. The equipment and procedures have been described in detail elsewhere (1). Values for the number, weight, z and z+1 average $\overline{DP}$'s were calculated from the chromatograms using a calibration curve based on the $\overline{DP}$ of cellulose samples determined viscosimetrically. It is pointed out that this procedure does not give absolute values of the DP's but does measure the relative changes resulting from the various treatments to which the samples have been subjected.

Results and Discussion

A comparison of the properties for round all-skin and all-core fibers is given in Table 2.

Table 2 Some Properties of All-skin and All-core Samples

	Tc	Ec	M	Tw	Ew	SM	Flex	H_2O Retn	H_2O Reg	$\overline{DP}$
Skin	4.58	10.9	127	3.10	28.5	2.82	6055	79.0	12.4	760
Core	2.71	4.9	126	1.21	11.7	8.36	423	86.2	11.3	735

It has been established that skin contains numerous small crystallites and the core fewer and larger crystallites (5). The same relationship holds for the amorphous or low order regions in these structures. Since the average cellulose molecular chain length (DP) is similar for both samples in Table 2, the molecules will pass through many more ordered regions in the all-skin fiber. The greater strength of the all-skin fiber is thus attributed to the number of ordered regions with which each cellulose molecule is associated. The arrangement of cellulose molecules between numerous small ordered and disordered regions produces a structure that is more extensible than the core. Not only are the smaller crystallites more able to slip past one another, because with relatively small contact surface area, there will be fewer hydrogen bonds; but ties between the crystallites will be less likely to restrict their movement relative to their length.

Skin swells less than core. Because the cellulose molecules are fixed in the crystallites at more frequent intervals, there will be less freedom of the small amorphous areas to swell. Since it is in the more open amorphous regions of the structure that water will be retained on centrifuging, water imbibition is lower in skin than for the core. On the other hand, moisture regain is higher. This is explained by an increased number of hydroxyl groups available for bonding with water as a result of the larger total surface area of the more numerous smaller crystallites.

The ratio of the wet to conditioned tenacity has been used to estimate the proportion of accessible hydrogen bonds in the amorphous regions (6). If the initial structures have similar molecular orientation and crystallinity, then this ratio will also qualitatively reflect the relative amount of skin and core in the structure. In the wet state, swelling disrupts the structure at the surface of the ordered regions and causes chain ends to become detached reducing the number of ties between the crystallites. Core swells more than skin and since there are fewer larger ordered regions, the reduction in tenacity is much greater than for skin. The water imbibition and the moisture regain for the samples used in the study are shown as a function of Tw/Tc in Figure 1. The less skin, the lower Tw/Tc and moisture regain and the greater water imbibition.

A rather striking observation that was made concerned the staining characteristics of the fibers. After 16 hours heating, every sample stained as all-skin. Stained cross sections for four of the samples are shown in Figure 2. The change in staining characteristics was related to the heating time with skin formation progressing from the outside of the filament. A first reaction to this observation was that an all-skin fiber could be made from an all-core fiber simply by heating. But on examination of the mechanical properties, it became evident that the transformation involved more than a simple conversion of core to skin. The ratio of wet to conditioned tenacity was used as a measure of the skin-core ratio. The effect of heating on this measurement for all-skin and all-core fibers is given in Table 3. Heating

Table 3 Heating; Effect on the Wet to Conditioned Tenacity Ratio

Hours at 184°C	Tw/Tc					
	All Skin			All Core		
	1	2	3	1	2	3
0	.677	.653	.605	.445	.458	.446
1	.642	.628	.502	.444	.470	.465
3	.591	.602	.428	.375	.461	.422
16	.516	.403	.389	.384	.380	.324

caused a decrease in the ratio for both fiber types, contrary to the expected increase if core were being

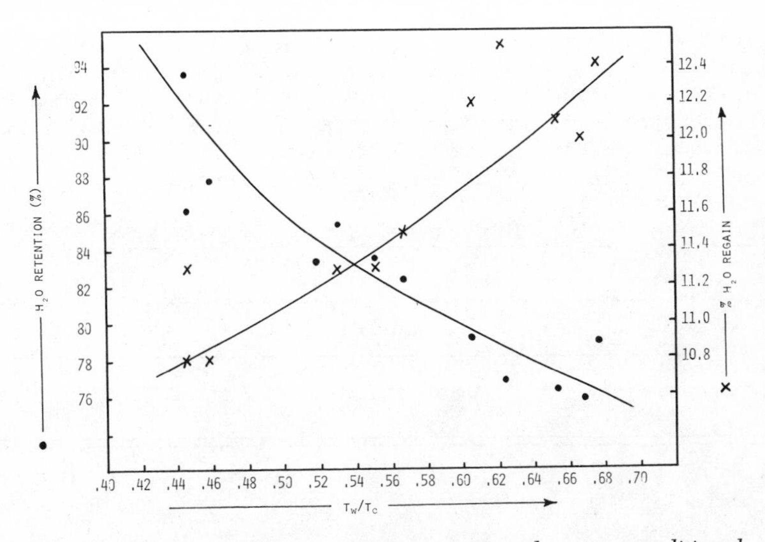

Figure 1. Water imbibition, moisture regain and wet to conditioned tenacity ratio

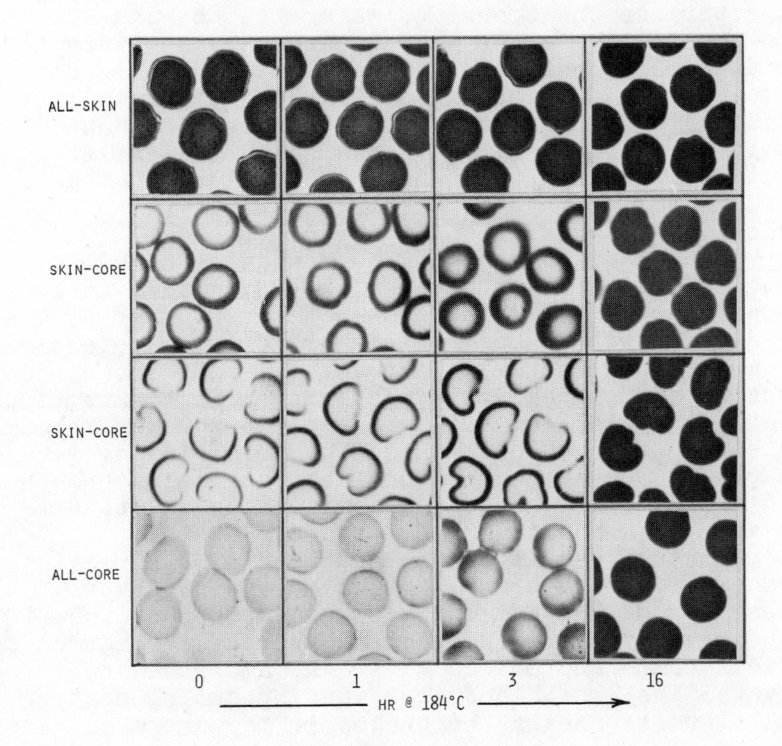

Figure 2. Stained cross sections

converted to skin. Initially, the ratio was much greater for the all-skin fiber, but after heating for 16 hours, the ratios were similar for both types, the most degradation having occurred in the all-skin fiber.

The water imbibition for these same samples is given in Table 4. The measurement is made by

Table 4 Heating; Effect on Water Imbibition

Hours at 184°C	% H_2O Retention					
	All Skin			All Core		
	1	2	3	1	2	3
0	79.0	76.3	79.2	93.6	87.7	86.2
1	76.8	76.8	77.8	81.0	79.2	85.3
3	74.9	81.2	80.8	74.6	69.2	77.8
16	70.5	74.0	75.6	61.8	61.6	61.9

centrifuging the wet sample under standard conditions to determine the amount of water held by the structure. The crystalline areas will have a very low water retention; association between cellulose molecules in these ordered regions will limit the accessibility and swelling. It is in the disordered or amorphous regions, where swelling is possible, that moisture will be retained. The water imbibition can thus be used as an estimate of the crystallinity with low values corresponding to higher order.

Because of the inclusion of a large number of small microscopic voids, the skin structure is less dense than core. This is one of the reasons it swells less and shrinks more. A higher moisture retention, such as shown initially for the all-core fiber, is usually associated with greater swelling. But, after heating, the all-core fibers exhibit a lower water retention than the all-skin fibers. This suggests that structural changes such as crystallite growth and crosslinking can occur more easily and to a greater extent with an all-core structure.

Moisture regain on the other hand measures the number of free hydroxyl groups in the structure. These can be the hydroxyl groups exposed on the surface of the crystallite and hydroxyl groups on the "tie" chains and molecules in the amorphous regions. Crystallite growth and crosslinking during heating will involve mostly the chains in the amorphous areas and cause a reduction in the moisture regain. The extent of this reduction will depend on the size and number of crystallites in the original structure.

Table 5 Heating; Effect on Moisture Regain

Hours at 108°C	Moisture Regain (%)					
	All Skin			All Core		
	1	2	3	1	2	3
0	12.4	12.1	12.2	10.8	10.8	11.3
1	12.1	13.5	12.1	10.5	10.4	10.6
3	11.9	11.6	10.9	10.2	10.1	10.6
16	11.2	10.6	10.9	9.4	10.4	9.7

The moisture regain data given in Table 5 is consistent with this interpretation. In the all-skin structure, the crystallites are smaller but more numerous. There is less material available in the amorphous areas and therefore crystallite growth on heating is restricted and a crosslinked structure containing many small voids will be formed. Under the same conditions, an all-core structure which contains fewer but larger ordered and disordered regions will form a more compact structure by crystallization and crosslinking. The shrinkage that occurred during heating in some cases caused the all-core fibers to break.

Accompanying these structural changes caused by heating, the tenacity and extensibility in both conditioned and wet states decrease. Since crystallite growth and crosslinking increase the strength, these observations suggest that other processes such as molecular degradation have also occurred. The area beneath the stress-strain curve is reduced and the breaking energy or toughness of the fiber is lowered. A consequence of this is reduced single fiber flex-life; with increased heating, both all-skin and all-core fibers become brittle. This is shown in Table 6.

Table 6 Heating; Effect on Single Fiber Flex Life

Hours at 184°C	Flex Life (Cycles					
	All Skin			All Core		
	1	2	3	1	2	3
0	6055	3750	5570	250	113	423
1	3037	4488	4687	102	Br	95
3	927	1210	4018	8	8	3
16	63	20	Br	Br	Br	Br

Heating caused a substantial reduction in the degree of polymerization ($\overline{DP}$). The measurements given in Table 7 were determined viscosimetrically using copper ethylenediamine as a solvent. The relationship between the time of heating and the reciprocal $\overline{DP}$ was

Table 7 Heating: Effect on Basic Degree of Polymerization (Viscosity Method)

Hours at 184°C	$\overline{DP}$					
	All Skin			All Core		
	1	2	3	1	2	3
0	760	790	720	580	600	735
1	580	580	570	515	525	605
3	440	460	425	395	405	460
16	175	165	155	150	160	140
Rate $Sec^{-1} x 10^7$	1.53	1.60	1.74	1.75	1.61	2.04

linear indicating that the degradation was random and obeyed first order kinetics. The rates shown in this table were calculated from the equation:

$$\frac{1}{DP_{(t)}} - \frac{1}{DP_{(o)}} = k \frac{t}{2}$$

They appear to be reasonably independent of the structure and degree of polymerization of the original samples. This is in agreement with earlier observations that irradiation depolymerization occurs equally in regions of low and high order.

During the initial heating, water was driven from the sample. The structure was annealed probably by a process of crystallization and the strength increased with some loss in extensibility. This is a characteristic of all rayon fibers, the strength increasing and extensibility decreasing as water is removed from the structure. At high temperature, the presence of moisture will cause appreciable hydrolytic degradation. Under the conditions used in these experiments, heating at 184°C in an air oven for up to 16 hours, degradation occurred as evidenced by the yellow/brown color of the samples after heating.

The weight loss, estimated from denier measurements, was less than 2% after 3 hours at 184°C for both core and skin fibers. After 16 hours heating, a 5% weight loss was estimated for the all-skin fiber but the all-core fiber was extremely brittle and denier measurements could not be made.

The weight average degree of polymerization, $\bar{M}w$, from GPC measurements is given in Table 8.

Table 8 Heating, Effect on Weight Average Degree of Polymerization (GPC)

Hours at 184°C	$\bar{M}w$, (100 ($\bar{M}w$ - $\overline{DP}$)/$\overline{DP}$)					
	All Skin			All Core		
	1	2	3	1	2	3
0	1234 (62)	1489 (88)	1176 (63)	618 (7)	720 (20)	873 (19)
1	1059 (83)	1344 (132)	1048 (84)	673 (31)	760 (45)	915 (57)
3	1026 (133)	1155 (151)	976 (130)	720 (80)	924 (128)	876 (90)
16	311 (78)	495 (200)	410 (164)	256 (71)	345 (116)	241 (72)

These values are all greater than the basic degree of polymerization, $\overline{DP}$ (Table 7) by the percentage shown in parenthesis. This can be attributed to several causes. Thus, the $\overline{DP}$ was obtained on each sample by a viscometric method. Theoretically, viscosity molecular weight data is only valid for distributions which are Gaussian. The GPC data which is calculated from the measured distribution is not sensitive to this parameter. Consequently, only with a Gaussian distribution will $\overline{M}w$ and $\overline{DP}$ be the same.

In other work, it has been shown that electron beam irradiation of cellulose yields a product susceptible to alkaline degradation. An additional depolymerization amounting to about a 10% $\overline{DP}$ loss on dissolving the irradiated product in alkaline cuene could be prevented by sodium borohydride reduction. It is probable that the extent of alkaline degradation will be influenced by the number of carbonyl and carboxyl groups formed during irradiation. Differences in the viscometric and GPC data are to be expected if the samples contain structural units that are resistant to acid hydrolysis and unstable in alkali.

For the unheated samples, the differences between $\overline{M}w$ and $\overline{DP}$ for the all-core fibers are much smaller than for the all-skin fibers (Table 8). This suggests that solubilizing the samples by nitration is influenced by crystallite size and number which are known to be different for skin and core structures. Incomplete separation of the molecules from the crystallites during dissolution would result in high $\overline{M}w$ values. The extent to which this can be caused by poor nitration techniques or by the presence of impurities in the sample has yet to be clarified.

The GPC data showed a different response to heating for the all-skin and all-core fibers. In the three all-skin samples, $\overline{M}w$ decreased on heating. With the all-core fibers, $\overline{M}w$ increased during the first few hours and then decreased. These different responses appear to be the result of the amount of material in the amorphous areas of the original structure available for crystallite growth, crosslinking, and other structural changes.

In all cases, heating increased the difference between $\overline{M}w$ and $\overline{DP}$ with the difference, as a percentage of the $\overline{DP}$, being greatest for the all-skin fibers. High $\overline{M}w$ values are to be expected if the solution of nitrated cellulose injected into the chromatograph contains crystallite fragments or gels. In some cases, the larger particles would not be fractionated in the chromatograph, the pore sizes in the column packing having been chosen to give the best fractionation for discrete molecules. This would result in the appearance of a "prehump" or peak in the chromatogram preceding elution of the fractionated material (7). Particles smaller than the largest pore size would be fractionated as large molecules. Prehumps were not observed for any of the samples, indicating that any structural residues in the solutions were being fractionated.

Table 9 Heating; Effect on DP Distribution

Hours at 184°C	DP Distribution									
	All Skin					All Core				
	$\overline{DP}$	$\bar{M}n$	$\bar{M}w$	$\bar{M}z$	$\bar{M}z+1$	$\overline{DP}$	$\bar{M}n$	$\bar{M}w$	$\bar{M}z$	$\bar{M}z+1$
0	760	422	1234	3723	7590	735	406	873	1963	4647
1	580	355	1059	3741	8042	605	354	915	2741	6635
3	440	311	1026	4078	8845	460	305	876	3292	7872
16	175	119	311	868	2177	140	107	241	479	832

The DP distributions calculated as the first moments of the particular distribution functions such as number, weight, z and z+1 distributions for an all-skin and an all-core fiber of similar original $\overline{DP}$ are compared in Table 9. It can be shown theoretically that the number average, $\bar{M}n$, is insensitive to changes in the number of large molecules. This, indeed, is the observation that can be made on these results, $\bar{M}n$ being similar for both structural types.

The original samples, before heating, stained as all-skin and all-core. After 16 hours heating causing degradation and structural changes, the samples both stained as all-skin. The change of $\bar{M}z$ and $\bar{M}z+1$ on heating indicates differences in the number of large molecules and particles in the sample. On the one hand, the more numerous and smaller crystallites of the all-skin fiber restrict the structural changes and very little change is observed in $\bar{M}z$ and $\bar{M}z+1$ before degradation. On the other hand, the larger amorphous regions in the all-core fiber allow extensive structural changes and $\bar{M}z$ and $\bar{M}z+1$ increase close to the values observed in the all-skin fiber before degradation.

Conclusions

Changes in the molecular weight distribution on heating a series of all-core and all-skin rayon fibers in the presence of air indicated that structural re-arrangements occurred during thermal degradation. The extent of structural rearrangement was related to the crystallite size and number and to the amount of material in the amorphous areas of the original structures. For all-core fibers which have more extensive disordered regions than all-skin fibers, the cross sections stained as all-skin after heating. Chemical and mechanical properties of the fibers were also affected by heating in agreement with the indicated structural changes. The rate of degradation calculated from the $\overline{DP}$ loss was similar for both fiber types.

Large differences between the basic degree of polymerization measured viscosimetrically and the weight average DP from GPC were attributed to deviation from a Gaussian distribution of molecular weight, to the presence of structural units that were resistant to acid hydrolysis and unstable in alkali and to incomplete separation of the molecules from the crystallites or structures during dissolution of the nitrated cellulose sample.

Literature Cited

1. Phifer, L.H., and Dyer, J., Separ. Sci. 6, 73 (1971)
2. Dyer, J., and Phifer, L.H., Separ. Sci., 6, 89 (1971)
3. Dyer, John and Phifer, Lyle H., J. Poly. Sci. C, #36, 103 (1971)
4. Morehead, F. F., ASTM Bull. 163, 54 (1950)
5. Sisson, Wayne A., Textile Res. J., 30, 153 (1960)
6. Bingham, B.E.M., Makromol. Chem., 77, 139 (1964)
7. Tanghe, L.J., Rebel, W.J., and Brewer, R.J., J. Poly. Sci. A-1, 8, 2935 (1970)

INDEX

D

E

W

X

Y

Z